TRUFFLE

Edible

Series Editor: Andrew F. Smith

EDIBLE is a revolutionary series of books dedicated to food and drink that explores the rich history of cuisine. Each book reveals the global history and culture of one type of food or beverage.

Already published

Apple Erika Janik   *Barbecue* Jonathan Deutsch and Megan J. Elias   *Beef* Lorna Piatti-Farnell   *Beer* Gavin D. Smith   *Brandy* Becky Sue Epstein   *Bread* William Rubel   *Cake* Nicola Humble   *Caviar* Nichola Fletcher   *Champagne* Becky Sue Epstein   *Cheese* Andrew Dalby   *Chocolate* Sarah Moss and Alexander Badenoch   *Cocktails* Joseph M. Carlin   *Curry* Colleen Taylor Sen   *Dates* Nawal Nasrallah   *Dumplings* Barbara Gallani   *Eggs* Diane Toops   *Figs* David C. Sutton   *Game* Paula Young Lee   *Gin* Lesley Jacobs Solmonson   *Hamburger* Andrew F. Smith   *Herbs* Gary Allen   *Hot Dog* Bruce Kraig   *Ice Cream* Laura B. Weiss   *Lemon* Toby Sonneman   *Lobster* Elisabeth Townsend   *Milk* Hannah Velten   *Mushroom* Cynthia D. Bertelsen   *Nuts* Ken Albala   *Offal* Nina Edwards   *Olive* Fabrizia Lanza   *Oranges* Clarissa Hyman   *Pancake* Ken Albala   *Pie* Janet Clarkson   *Pineapple* Kaori O' Connor   *Pizza* Carol Helstosky   *Pork* Katharine M. Rogers   *Potato* Andrew F. Smith   *Pudding* Jeri Quinzio   *Rice* Renee Marton   *Rum* Richard Foss   *Salmon* Nicolaas Mink   *Sandwich* Bee Wilson   *Sauces* Maryann Tebben   *Soup* Janet Clarkson   *Spices* Fred Czarra   *Sugar* Andrew F. Smith   *Tea* Helen Saberi   *Tequila* Ian Williams   *Truffle* Zachary Nowak   *Vodka* Patricia Herlihy   *Whiskey* Kevin R. Kosar   *Wine* Marc Millon

Truffle

A Global History

Zachary Nowak

REAKTION BOOKS

For my mentor and friend
Simon Young

Published by Reaktion Books Ltd
33 Great Sutton Street
London EC1V 0DX, UK
www.reaktionbooks.co.uk

First published 2015

Printed and bound in China

A catalogue record for this book is available
from the British Library

ISBN 978 1 78023 436 6

Contents

Introduction: A Shy Fungus, the Jewel of Cuisine

Like a platypus or a Venus flytrap, a truffle seems not to fit neatly into any category, a fact that has led to much confusion through the centuries. The ancient Roman naturalist Pliny, in his encyclopaedia entry on truffles in the first century CE, noted that

> among the most wonderful of all things is the fact that something can spring up and live without a root. When there have been showers in autumn and frequent thunder-storms, truffles are produced, thunder contributing more particularly to this development. Now whether this imper-fection of the earth – for it cannot be said to be anything else – grows, or whether it lives or not, are questions which I think cannot be easily explained.

A sixteenth-century German herbalist, Hieronymus Bock, gave a similar and equally erroneous explanation of where truffles came from: 'Truffles are neither herbs nor roots, nor flowers, nor seeds, but simply superfluous moisture of the earth, of trees, or rotten wood.' By the early nineteenth cen-tury, truffles were at least better appreciated, if still not fully understood: the famous French gourmand Jean Anthelme

Summer black truffle.

Brillat-Savarin called them 'the jewel of cuisine'. The composer Rossini apparently agreed, referring to truffles as 'the Mozart of mushrooms'.

A truffle is a kind of mushroom, which is technically the fruiting body of a spore. Although they do not use photosynthesis to make their own food, most mushrooms are similar to plants, living both above and below ground. They have an outer covering called a peridium and an interior flesh called gleba; these come in many different colours and textures, and those differences are the first method of distinguishing the various species. All truffles are technically the 'reproductive body' or 'fruiting body' that produces spores. These spores are contained in tiny sacs called asci, and can vary quite a lot in appearance (they can be circular or oval, spiked or smooth, and so on). Microscopic examination is the second step in the identification of truffles: close up, truffles that have very similar peridium and gleba might differ greatly in the number of spores in their asci, and in the appearance of the spores themselves.

When carried away by the wind or by animals, these spores produce new mushrooms. While most mushrooms have their 'fruiting body' (the toadstool) above ground, truffles are reclusive mushrooms, hiding theirs (the truffle) underground. In common with some other fungi, truffles make a biological deal with the tree among whose roots they grow. While some mycologists (scientists who study fungi) see the trade as less than amicable, the tree ultimately 'allows' the truffle to wrap small hairs around its roots; think of a kind of loose-knit glove. The tree provides the truffle with sugars and other nutrients through this cellular glove, called a Hartig net. In return the truffle, which forms a huge network of small hairs called *hyphae*, absorbs water and minerals for the tree to use, greatly expanding the reach of the tree's root system. This arrangement allows the truffle to do without the effort of photosynthesis and to hide for most of its life from unwanted attention.

Every organism has to reproduce itself and expand its habitat, however, and many plants rely on animals for help with reproduction. In a deal akin to that between truffles and trees, many flowering plants offer nectar to bees. When the bees crawl into the flower to collect the nectar, they get pollen on their legs, which they carry to the next flower, pollinating it. Some plants go even further, providing a sugary morsel (a fruit) to an animal in return for transporting its seed to another location. The seed, which passes through the animal's gut, is deposited far away, packaged in a helpful coating of organic fertilizer.

Truffles, too, rely on animals to reproduce. The strong smell we associate with truffles is their perfume: when their spores are ripe, truffles produce these aromas to attract 'fungivores'. These fungus-eaters dig up the truffle and eat it, depositing the spores several hours later in another spot in the woods. Fungivores include pigs – which are no longer much used to find the hidden fungi, despite pop-culture pairings of swine and truffles – and dogs, which are easier to train to leave the truffles to the truffle-hunter and eat a doggie treat instead. If dogs are a man's best friend, a truffle's best friend is *Homo sapiens*. Foreign truffles now inhabit continents where they were not found a hundred years ago, and it is humans who have made that possible.

Humans, like other fungivores, are interested in truffles because they smell good. As we will see, truffles went from being just another (albeit odd) mushroom to being a symbol of haute cuisine. Their perfume makes them ideal for transforming a quotidian recipe into a remarkable dish, and their relative scarcity ensures that their price remains high. As the truffle's aroma imbues other ingredients with a delicious taste,

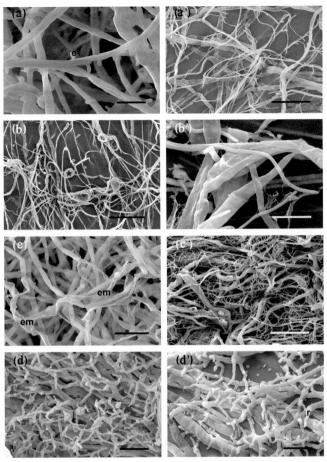

Scanning electron microscope photos of the hyphae of the *bianchetto* truffle (*Tuber borchii*).

its exclusivity gives its privileged consumers yet another means of distancing themselves from the crowd.

Just as history books often revolve around epic battles and prominent leaders, books about truffles often focus on the two superstars, the Périgord black truffle (*Tuber melanosporum*)

Truffle dog Sole at work.

from France and the Alba white truffle (*T. magnatum*) from Italy. A truly global history of truffles must include the whole family, however. The following chapters will introduce the reader to these other members, whose stories show how Europe's culinary culture changed through the centuries, as it encountered the New World and the Developing World.

The French, when talking about truffles, often refer to *la grande mystique*. Truffles' mystique once revolved around the questions 'What are truffles? How do they grow? Why can't they be cultivated?' Thanks to modern science, most of these questions have been answered. Truffles are the hypogeous (underground) fruiting bodies of certain fungi; they mature spores and attract fungivores to disperse them; and some kinds of truffle can now be cultivated. The questions we ask about truffles have changed; our principal interrogative is now 'What should truffles be?' Should truffles be 'all natural', gathered by old men with dogs in the woods in late autumn, or should we try to cultivate them? If they should be natural, then they will remain largely on the tables of our culinary elite. Should a truffle be defined – legally – as one of the European species, or can North American and Asian hypogeous fungal fruiting bodies carry the moniker 'truffle'? How did a homely, halfway mushroom come to be a marker of culinary distinction and a symbol, simultaneously, of our rural past and our globalized future? *Truffle: A Global History* will illuminate each side of these subjects.

I

Truffles in the Sand

The recorded history of truffles begins in a place where it seems unlikely that one would find underground fungi: the desert. From these earliest mentions of the truffle, we can see the regard – both high and low – in which these mysterious foodstuffs were held. Truffles were important enough to arouse strong emotions in the rulers of men, but they were at the same time a symbol of the difference between 'barbarians' and their 'civilized' neighbours.

In a dusty corner of Syria about 50 km (30 miles) north of the border with Iraq, there is a large mound on the eastern bank of the Euphrates River. Archaeologists get excited about mounds in otherwise flat landscapes, since often a little digging reveals the crumbly bits of an ancient empire; they call these promising mounds 'tells'.

Excavation of this particular mound, Tell Hariri, began in the 1920s after local Bedouins discovered a headless statue. The French, then governing Syria, uncovered the remains of a large city that appeared to have been methodically emptied and then burned. Among the rubble was a huge cache of clay tablets inscribed with a stylus in the ancient Akkadian language. By the late 1930s most of the tablets had been transcribed and translated. The city, called Mari, had been the capital of a small kingdom. The inhabitants were well known in the ancient Near Eastern world: the Sumerians gave them the name Martu, while the Israelites referred to their neighbours as Amori. The Amorites, as modern archaeologists call them, were a rambunctious bunch. Initially nomadic warriors, they later settled down in territory around Mount Jebel Bishri. Sedentary living did not, however, make them less bellicose: they constantly attacked their neighbours, including the powerful Babylonians further down the Euphrates.

The Amorites' last king, who ruled from 1780 to 1760 BCE, was named Zimri-Lim, and he figures in many of the clay tablets. Preserved in baked clay we find his correspondence regarding the typical minutiae of the busy bureaucracy of an ancient empire: taxes due, census-taking in outlying areas, and relationships with friends and foes. But the store of letters also reveals a more human king. We find his order to increase his queen's wine allowance, his request to his personal diviner to interpret a troubling dream about his powerful ally, Hammurabi of Babylon, and a complaint about truffles.

On a clay tablet catalogued as xiv.35, we read a response by the bureaucrat Yaqqim-Addu to what must have been an angry letter from Zimri-Lim:

> Ever since I reached Saggaratum five days ago, I have continuously dispatched truffles to my lord. But my lord has written to me: 'You have sent me bad truffles!' But my lord ought not to condemn with regards to the truffles. I have sent my lord what they have picked for me.

Zimri-Lim was apparently a connoisseur of truffles, since there is yet another mention of the underground fungi among the clay tablets found in the ruined palace: Kibri-Dagan, the governor of the city of Terqa, wrote:

Now I am having those truffles sent to my lord . . . the one case of truffles and the one tablet which [personal name] sent me, now I am sending the case and the tablet which they brought me on to my lord, both under seal.

Perhaps the king should have spent more time on diplomacy and less on gastronomy, though. Some time around 1760 BCE his erstwhile ally Hammurabi turned against him; this had been foreseen by Zimri-Lim's diviners, who had prophesied Babylonian treachery. Despite their number, there are many questions that the smoke-stained tablets cannot answer. One is about the king's fate after Hammurabi conquered Mari, although one modern commentator euphemistically guesses that it was 'unenviable'. Another unanswered question is

Bedouin woman and boy digging *kimme* truffles in Syria, 1939.

what the king's last meal was the day before his capital fell; given Zimri-Lim's predilection for them, perhaps the menu included truffles.

For the victors, the result of this war with the Amorites was divine justice, a view in which the truffle figures as a mark of barbarity. The Babylonians, whose empire was on the rich farmland between the Tigris and the Euphrates, had long viewed the Amorites with suspicion, as uncouth barbarians who had long rejected agriculture; for the Babylonians, that was rejecting civilization itself. Fears of the other had been transmuted into a myth called 'The Marriage of Martu'. In the myth, a Babylonian girl named Adjar-Kidug contemplates marrying Martu, the main god of the Amorites. Adjar-Kidug's friend tries to dissuade her, calling the Amorites 'monkeys' and noting that they live in tents on the mountains. The worst accusation, though, is that Martu (and by extension, the Amorites) is not civilized in his eating habits: 'He lives in the mountains and ignores the places of gods, digs up truffles in the foothills, does not know how to bend the knee [do agriculture], and eats raw meat. My friend, why would you marry Martu?'

Becoming a sedentary farmer in the ancient world was the equivalent of being civilized, and being civilized meant hating (and fearing) people whose food did not all come from a field or a garden. This distrust of hunter-gatherers, pastoralists and other nomadic people was common in the ancient world and led to edible wild food – such as truffles – being viewed with ambivalence at best and suspicion at worst.

Bread-eaters and Sandy Truffles

The ancient Greeks shared this idea of civilization being what separated men from beasts: culinary artifice, not naturalness, was highly valued. The best example of this was bread: wheat had to be sowed, tended, harvested and transformed into bread, the mark of a man. For Homer, 'bread-eater' was synonymous with 'man'. Relying on root vegetables and wild plants for sustenance was a sign of great poverty. Yet while ideology for the Greeks focused culinary attention on the artificial, the poor still took advantage of naturally occurring foodstuffs, such as truffles. The Greeks were perhaps the originators of some of the more persistent beliefs about truffles. One was their use as an aphrodisiac. The writer Philoxenus of Leucas, in his *Symposium* of the fifth century BCE, maintains that truffles baked in embers are conducive to 'amorous play'. Other ancient beliefs show great differences between their perspective on the culinary place of the truffle and even what a truffle was, compared to later understanding.

One of the most literate peoples of the ancient world, the Greeks were the inventors of many genres, including the encyclopaedia. In his botanical work *Enquiry into Plants*, Theophrastus (371–287 BCE) called truffles 'a natural phenomenon of great complexity, one of the strangest plants, without root, stem, fibre, branch, bud, leaf or flower', noting that there were a number of different kinds of truffle, including one he called *misu*, which grew near Cyrene in present-day Libya, or 'wherever the ground is sandy'. Talking about their generation, he supposes they might have a seed, noting that in any event the more thunder there is in autumn, the more plentifully they grow. As we will see, thunder as a catalyst to truffle fruiting was a commonly held belief in the ancient world.

Not all Greek authors were as neutral in their descriptions. Nicander, writing in about 185 BCE, says truffles are 'the evil ferment of the earth that men generally call by the name of fungus'. Dioscorides, who wrote a work in the first century CE on the use of plants in medicine (*De materia medica*), left us

Statue of Theophrastus in the Palermo Botanical Gardens.

with a short entry on truffles that seems to refer to another plant. Calling the truffle *hydnon*, Dioscorides follows Theophrastus in putting it in a category of smooth-skinned flora. He further indicates that it is 'a round, pale, yellow root without leaves or stalk. It is dug up in the spring and is edible eaten either raw or boiled.' To modern readers, this is a particularly curious description. While there are white truffles, it is quite a stretch to describe them as round or yellow. Truffles are mostly dark and they are only vaguely round, with awkward bumps that make them difficult to peel.

Yet the physical descriptions of Roman writers resemble those of the Greeks. Pliny the Elder was the most intrepid

A vendor offering desert truffles (*Terfezia arenaria*) in Kuwait.

Roman naturalist. He wrote a monumental work called *Natural History*, whose 37 volumes contained all there was to know about the natural world in the first century CE. Pliny's entry on truffles is indicative of the confusion that ancient writers had classifying this strange underground 'swelling', the meaning of the Latin word that Pliny uses, *tuber*. The entry – after the passage we have already seen about how thunder in autumn produces more truffles – mentions a kind of truffle that is 'full of sand'. The author recounts an anecdote that suggests the origin of truffles; one of Pliny's friends had bitten into a truffle and found a *denarius* (a Roman coin) in the middle of it, which all but broke his front teeth. Pliny suggests that this is 'evident proof that the truffle is nothing else but an agglomeration of elementary earth'.

What strikes modern readers about Pliny's comments on truffles, other than the description of the truffle as smooth and yellow, is what he writes about the fungi's provenance. He does not give the mountains of the Italian peninsula as the truffle's habitat, but rather names 'Africa', which for the Romans meant the north African littoral, present-day Libya and Tunisia. This is echoed by a later writer, Juvenal, in one of his satires. Juvenal has a glutton at a Roman feast, referring to northern Africa's role in providing much of Rome's grain at the time, cry: 'Keep your grain to yourself, O Libya! Unyoke your oxen, if only you send us truffles!'

The explanation for these seemingly erroneous descriptions – truffles that were smooth and yellow and grew in the sand, instead of knobbly, brown and growing in moist forest soil – is that ancient writers from Zimri-Lim to Juvenal were describing another botanical genus. The truffles one finds in European markets today – all from the genus *Tuber* – are not from the Mediterranean's sandy eastern and southern rims, but rather from the wetter northern side of what the Romans

Mosaic, possibly from the 4th century CE, from the ruined Roman city of Sabratha (now in Libya). Despite the green leaf – probably an artistic flourish – the food represented here seems to be the desert truffle.

called *Mare nostrum*, 'our sea', an indication of how integrated the whole Mediterranean basin was.

The truffles of the classical world were mainly from the genus *Terfezia*: these are similar to their more famous cousins but much more abundant, and so cost much less. While they are now practically unknown in Europe, they are still widely eaten in the Muslim world, from Morocco all the way to the Arabian peninsula. While the species in the *Terfezia* genus do not grow in forests, they have a similar symbiotic relationship to that of the *Tuber* species, although with plants other than trees. One of the main companions of desert truffles is the rock rose, and it is under this plant that present-day north African and Middle Eastern truffle-hunters find their truffles. A telltale sign is not only the rock rose itself, but a crack in the dry earth around it, a sign of a truffle swelling in the sand below. Anyone who samples desert truffles today will see how accurate the ancients' descriptions were: desert truffles are much smoother and more uniform than their northern cousins, and are a deep yellow. They often have sand inside them, which necessitates a thorough washing before preparation. While examples of *Tuber magnatum* (the fabled Alba white truffle) can bring as much as $1,000 a pound, desert truffles sell for only $20 or $30 a pound. Their much weaker aroma makes them better as a vegetable than as a garnish.

The ancient Romans, even more than the Greeks, were prejudiced against peoples whose food came from natural resources rather than intentional cultivation. For the Romans, food geography was divided into three zones. The first was the city, called *civitas* in Latin: we can see from the similarities in the two words that the city for the Romans was the home of civilized living and eating. Beyond the city lay the *ager*, the fields, where peasants and slaves produced the so-called Mediterranean triad (wheat, grapes and olives to make bread, wine and olive oil), as well as market gardens for other crops.

Outside that ring of fields and gardens was the *saltus*, the wilderness. According to official Roman alimentary ideology, the forests and swamps were the domain of animals and beast-like humans, the barbarians who instead hunted game, fished and collected 'spontaneous vegetables' like truffles. Of course this ideology corresponded only partially to reality. Roman peasants often supplemented their diet with whatever they could find in the *saltus*, and some of these foodstuffs did make their way to the tables of the elite. Indeed, north African truffles were eaten by the rich, but they first had to be 'ennobled' by spices.

We think of heavily spiced food as medieval, but the long-distance import of spices into Europe began during the Roman Empire. Despite patriotic laments about the spices being a drain on the empire's reserves of silver – Pliny himself came up with the figure of 100 million *sesterces* a year – spices in general and pepper in particular were popular with those who could afford them. Roman millionaires could display their wealth at a banquet in several ways: by serving extremely rare foods, such as flamingo tongues and the womb of a spayed sow; or by adding exotic spices to otherwise plebeian food.

Pepper and Yellow Bile

To make sense of Roman culinary practice with regard to truffles, it is necessary to understand the role of spices and the theory of the humours in explaining human health. Today we use truffles as a garnish, to add flavour to a dish, but for the elite of the Roman Empire the desert truffle (product of the *saltus*) had to be 'civilized' with spices. The most famous Roman cookbook, *De re coquinaria* (*The Art of Cookery*), provides a glimpse into Roman imperial cuisine. The cookbook itself is something of a mystery: while the author is usually given as 'Apicius', this is more shorthand than fact. Marcus Gavius Apicius was a Roman nobleman and gourmand of the first century CE, but the list of recipes is more likely the work of an anonymous author in the late fourth or early fifth century CE.

The Art of Cookery gives six recipes for truffles. The first, entitled simply *Tubera* ('Truffles'), instructs the cook to scrape the truffles, parboil them, put them on a skewer and half-fry them: 'Then place them in a saucepan with oil, broth, reduced wine, wine, pepper and honey. When done put the truffles aside, thicken the broth with a roux, decorate the truffles nicely, and serve.' We can tell that this recipe is intended for wealthy households not only because of the use of pepper –

an extremely expensive item at the time, and found in more than 85 per cent of Apicius' recipes – but also because of the elaborate preparation needed. While we think of fast food as a relatively recent culinary corruption, most poor Romans ate out, buying fried honey cakes and sausages from corner shops. These eat-and-go establishments are still visible in the ruins of the Roman city of Pompeii, where shops called *thermopolium* on street corners had large amphorae containing food set into counters. Very few Romans could afford a kitchen where a cook could reduce wine and fry truffles.

Depending on when *The Art of Cookery* was written, it may contain one of the last written references to truffles in Western Europe for centuries. In 380 CE one of the early Christian fathers, St Ambrose, wrote to a friend named Felix that he was amazed at the size of the truffles Felix had sent him. St Ambrose says that he was not embarrassed to show the truffles to friends, and while he gave some away, he kept a few for himself. He was writing at a time when Christianity was beginning to grow into the administrative network of the empire as truffles' hyphae grow into the roots of trees. The Christian Church incorporated some of the empire's trad-itions, placing the Mediterranean triad of grapes, wheat and olives at the centre of its liturgy: wine and bread became Christ's body and blood, and anointing was done with olive oil. While some of its food ideology survived, the empire itself did not: having pushed back the much larger attacks of the Germanic tribes in the third century, it succumbed to renewed invasions in the fifth. The formal end of the Western Roman Empire in 473 CE led to a change in European food culture, although some elements of the classical culinary tradition remained.

One of these was the theory that digestion was a sort of internal cooking, and that to prevent illness one had to maintain equilibrium in one's diet. The body was seen as containing four liquids or 'humours' – blood, phlegm, yellow and black bile – and too much of one kind of food could throw off their delicate balance and sicken the eater. Each humour had a characteristic: blood was hot and dry, for example, while black bile was cold and moist. The right combination of hot, cold, dry and moist foods would be absorbed into the body and would maintain the correct balance of the four humours. Foods could also be mixed to neutralize deleterious effects. Salted ham was hot and dry but could be combined with melon (cold and wet) to make it safe; this combination, incidentally, is still popular in Italy today, although not for humoral reasons.

It was essential for each person to know his personal humoral balance. In general, women were considered moist and cold, men hot and dry (although men became colder with age). A balanced meal, then, was different for each person. While this theory came out of the Hippocratic tradition in Greece, the main proponent in Rome was a physician named Claudius Galenus. We call him Galen, and he was born around 129 CE in what is now Turkey. A prolific writer, Galen wrote manuscripts that survived the fall of the empire and were part of the medical knowledge transmitted to late medieval Europe. In his book *On The Properties of Foodstuffs*, Galen weighs in on truffles:

> Although they have no obvious quality, one must also number these amongst the roots and bulbs. For that very reason those who use them do so as a base for season-ings, in the way that they use others which they call bland, harmless and watery in taste. It is a common feature of all these that not even the nutriment being distributed to the body has any singular property but, while it is rather cool, it is itself similar in thickness to what has been eaten

– thicker from truffles, but moister and thinner from the bitter cucumber, and along the same lines with the others.

Given the evaluation – bland, better as a base for flavour than a source – we know that Galen was referring to desert truffles, a species he must have encountered in his youth in what was then known as Asia Minor.

On 4 September in the year 476 CE Romulus Augustus, the last emperor of the Western Roman Empire, went before the Senate and abdicated. This was done at the 'request' of Odoacer, the barbarian soldier who, as soon as the abdication was accepted by the Senate, declared himself king. Little is written about Romulus Augustus' later life, but sources agree that he was probably pensioned off and sent to live in a sumptuous villa on an island in the Bay of Naples. The island – now a peninsula with the imposing fortress of Castello d'Ovo on it – was near the port of the city, and one is left to wonder if, as the ships carrying provisions from north Africa were fewer and fewer, Romulus missed his truffles.

2

The Fall and Rise of the Truffle

The so-called Dark Ages are often portrayed as a long period in which civilized life was destroyed by 'barbarian' tribes. To some extent, this was true: the magnificent buildings and roads the Romans had built fell into disrepair; literacy and literary production declined precipitously; and even gastronomy (limited as it now was to local products) was impoverished. There were upsides, however, especially for the common folk. There was no centralized administration to construct monumental buildings, but the tax load was greatly reduced. Spice imports slowed to a trickle, but such goods had in any event been out of reach of the commoner, and so their lack was hardly noticed. What might have been bad for the Roman elite was good for gastronomy, because the truffles that we now grate on to plates of pasta or use to make exquisite sauces were discovered after the empire fell.

The Mediterranean Triad Meets 'Barbarian' Food

The warriors who took control of what is now western Europe became the petty lords of the areas they held, and their descendants ('noble' by virtue of their ancestors' militarism) had none of the Roman conceit against the 'fruits of the woods'. The *saltus*, the area outside the cultivated fields, increased as what had been open land reverted to woods and marshes. While the Germanic tribes accepted some of the classical diet into their own – indeed, the Mediterranean triad became central to Christianity, as the consecrated Host and wine and the holy oil – they retained their appreciation for foods that had been foraged or hunted. One of these was the truffle: not the desert truffle so well known to the Romans, but the European truffles that we find in symbiosis with oaks, hazelnuts and other hardwoods.

There is unfortunately no record of the discovery of these truffles, but it is easy to imagine how it happened. While the Romans had certainly eaten pork and enjoyed cured hams and sausages, it was the Germanic tribes that took over the former empire who brought charcuterie to a higher level of appreciation. The wild pig (*Sus scrofa*) had been domesticated in the Middle East perhaps as early as 13,000 BCE. Its modern descendant, the domestic pig (*Sus scrofa domesticus*), is only a subspecies of the original wild ancestor, and can interbreed easily with the wild boar still present in Europe's forests. While we think of pigs as pink, relatively docile animals, the medieval pig was much closer to its wild cousins. This is easy to see when we look at pigs in medieval art: they are large, hairy animals with long tusks, much less portly than our present-day animals, and often dark red. A painting from the *Très Riches Heures* of the Duc de Berry, a book of hours written and illuminated in the fifteenth century, shows pigs as they had been represented for centuries. While they are 'domesticated' – they are, after all, accompanied by swineherds – they exist at the border of the civilized and the wild. Some pigs are still in the open, while others are already in the woods,

A 15th-century illumination of truffle hunting, from the *Très Riches Heures* of the Duc de Berry. Medieval pigs looked little like their modern counterparts, but both enjoy eating truffles.

eating acorns that the swineherd knocks from the trees with a stick.

Throughout the Middle Ages (and indeed until very recently in Mediterranean Europe), people who raised pigs let them roam freely or, as in the *Très Riches Heures* painting, partly supervised in the thick woods that covered much of the land. It is also interesting that in the early Middle Ages forests were measured not in acres but in the number of pigs they could feed. The pigs, omnivorous scavengers like their wild cousins, rooted around in the undergrowth for berries, insects, snakes and nuts, and dug for roots, tubers . . . and truffles. Their well-developed sense of smell made it easy for them to find mature truffles by the scent they emitted, and to dig them up and eat them. The only difference now was that their human 'supervisors' were with them in the woods. Whether truffles were 'discovered' by a particular anonymous yet attentive swineherd, or whether this happened many times all over Europe, is a mystery. It is a mystery, too, what led that swineherd to try to eat this smelly, grubby underground mushroom, but luckily he was as adventurous in the kitchen as he had been attentive to his pigs in the woods. Somehow the knowledge of finding truffles with pigs spread, as did their use as food.

Most conventional histories of truffles treat the period when European truffles were probably 'discovered' as a dark age, because no documents have been found in Christian Western Europe that specifically mention truffles between St Ambrose's letter and a twelfth-century poem. This blind spot of historical methodology – 'If it wasn't written down, did it really happen?' – leads us to another rhetorical question, which we can paraphrase as follows: 'If a swineherd finds a hypogeous fungi while alone in the woods, did he really discover truffle cuisine?'

Other cultures were discussing these delicacies from underground, however. The tenth-century Byzantine monk, writer and politician Michael Psellos mentioned truffles in one of the several hundred personal letters of his that we still have today. While it is unclear whether he is referring to the desert truffle or the European truffle, he thanks the emperor's brother for the gift of a basket of truffles, which he compares to the 'richest of lands'. The eleventh-century Persian physician Ibn Sina (known in the West as Avicenna) did not have a completely negative opinion of truffles, since he recommended them for healing wounds as well as treating weakness and vomiting; but he did note that they are 'apt to induce apoplexy and paralysis'. At the other end of the Muslim world, in the city he knew as Ishbiliya (Seville), the writer Ibn Abdun had a dimmer view of truffles. In a treatise on the government and public morals of the city, he recommended that the sale of truffles be banned because of their use as an aphrodisiac. The great Moses Maimonides, a Jewish commentator on the Torah, joined Ibn Abdun in blacklisting these fungi (along with other mushrooms) as something to be avoided by the person who wanted a long, healthy life.

Platina and Norcian Pigs

Although not everyone who wrote treatises (medical or political) was a fan, truffles did have enthusiastic supporters. One was none other than Petrarch, the Italian poet, who, although he did not understand that the 'Dark Ages' had been necessary for the discovery of the treasures hidden in Italy's woods, did seem to enjoy the culinary delight. Talking about how light alone cannot induce the earth to make truffles, Petrarch says in one of his sonnets (here translated by Gillian Riley):

Truffles and pasta: a dish fit for a king.

Nor that glow which lights up
Hills and dales with little flowers,
But cannot penetrate the earth,
Which, pregnant by itself alone,
Produces this fruit so rare.

According to theories prevailing at the time, truffles were simply a combination of two elements, earth and water. This idea of the origin of truffles was drawn from the Greek and Roman texts that mentioned these shy mushrooms. Since the 'rebirth of art and learning' meant for the Renaissance man the rebirth of *classical* art and learning, writers from the fifteenth century onwards who discussed truffles never failed to cite the ancient texts that they assumed were talking about the same truffles they were enjoying.

One of the most famous Renaissance gourmands was Bartolomeo Sacchi, better known as Platina (the name he invented for himself, based on the Latin name of his home town, Piadena in northern Italy). After receiving a humanist

Stefano della Bella (1610–1664), *Man with Pig*, drawing.

education (which at the time meant learning Latin and Greek), Platina arrived in Rome and won a position in the papal bureaucracy. A social climber, he quickly realized the importance of hobnobbing with prelates, and he was invited to spend the summer of 1463 at the villa of the patriarch of Aquileia. He got to know the patriarch's cook, Maestro Martino, whom he later acknowledged to be an incredible culinary improviser and the source of most of Platina's recipes. While Martino had written his own book, *The Art of Cooking*, it was Platina who made it appetizing to the educated readers of the day.

Platina copied about half of Martin's recipes, and added an elaborate scaffolding of citations from Roman and Greek authors. Written in Latin and published in 1474, Platina's *De honesta voluptate et valetudine* (On Respectable Pleasure and Good Health) is an odd halfway point between slavish devotion to the ancients and a growing confidence that the present could be greater than the past. It is precisely in the field of gastronomy that Platina makes a break with the intellectuals of the past: 'There is no reason why we should prefer our ancestors' tastes to our own. Even if they surpassed us in nearly all the arts, in taste alone we are undefeated.' Still, it is significant that while Platina's tastes may have been different from those of his Roman ancestors, they were very similar to those of his medieval antecedents. And while his taste may have been different from Pliny's, his way of thinking about food was not: reading the recipes in *De honesta voluptate et valetudine*, we see that despite the passage of more than a millennium since Galen, the humoral theory is still the main way of evaluating foods for their nutritional value.

Platina is perfunctory in his entry on truffles, and the careful reader can see that it was written after consulting ancient texts, not contemporary truffle-hunters. After the well-worn quotations from the Romans about thunder and rain, and truffles being harmful to the teeth, and references to the north coast of Africa, Platina remarks that the pigs in the Umbrian mountain city of Norcia are remarkable for their sense of smell. After finding the truffles, he claims, the pigs will back off and leave them for the farmer as soon as he tickles their ears. The truffles should be washed in wine (a common purifier for dangerous foods), cooked with pepper and served with meat.

Another book typical of this genre was Alfonso Ceccarelli's *De tuberibus* (On Truffles). Although some earlier men writing commentaries on Greek botanical texts had mentioned truffles, Ceccarelli's book (actually more of a short treatise) was the first work to have truffles as its subject. Ceccarelli was born in the town of Bevagna in 1532. Bevagna is in the Tiber valley but is only 8 km (5 miles) from where the Apennine mountains shoot up, and only about 65 km (40 miles) as the crow flies from the famous truffle-hunting areas between the Umbrian cities of Spoleto and Norcia. Ceccarelli's mother came from an illustrious, if minor, noble family, and his father was a notary. Ceccarelli and his seven siblings must have enjoyed truffles in their youth, and he carried this fond memory with him to Padua, where he studied medicine.

It was in Padua that Ceccarelli probably wrote his treatise on truffles. The book is made up of nineteen short chapters, each of which treats an individual question about truffles: if they can be sown like seeds (Chapter vi), in which season they should be 'extracted' (Chapter xi), and if they are roots, fruits or plants (Chapter xv), among others. Although truffle experts today confirm that it is in places surprisingly modern, the text is all in Latin and mostly follows the style of the day, with copious citations from Pliny, Dioscorides and other Roman and Greek writers who mention (or even seem to mention) truffles. Only occasionally will Ceccarelli, like a hesitant under-graduate, advance his own opinion – for example, when he considers whether truffles are found in the New World as well. What is unclear about the text, written as it is by a young man who was apparently quite a personality, is whether it is serious or simply a comic parody, meant to ape the tedious manuals that he and his classmates had to study for their exams. A close reading of the text reveals that Ceccarelli was not above cheating: to buttress a point about white truffles vis-à-vis black ones, he invents a quotation from Avicenna.

The Columbian Exchange

One of the texts that Ceccarelli probably had to read (or at least consult) for his medical degree was Pietro Andrea Mattioli's *Discorsi* (Discourses; published in 1544, when Ceccarelli was only twelve years old). Mattioli, who was from Siena, had also studied in Padua. His book quickly became a standard university text; while ostensibly a mere commentary on the translated botanical work of the Greek philosopher Dioscorides, it was in reality more ambitious, an attempt to incorporate the flood of new botanical discoveries into the received knowledge on the plant and animal world. The project was necessary because the 'discovery' of the New World had set in motion what historians refer to as the Columbian Exchange.

This phrase, coined in 1973 by the historian Alfred Crosby, refers to the massive exchange of food products (and intoxicants: think coffee and chocolate) between Europe and the Americas. For those who are not food historians, what is most striking is to imagine Eurasian food – or indeed most world cuisines – without these American products. Think of the menu of any restaurant within a kilometre of Berlin's Pergamon Museum, Istanbul's Hagia Sophia or Saigon's Reunification Palace, and imagine it without potatoes, tomatoes or chillies, not to mention maize, New World beans or chocolate. No potato soup, no kebabs filled with tomatoes, no stir-fries spiced with bird's-eye chillies. The voyages of discovery and imperialism undertaken by the Europeans in the late fifteenth and early sixteenth centuries set off the largest anthropogenic movement of flora and fauna in history – none of which, of course, had been described by the ancients.

The Renaissance was an interesting intellectual period for Europe, for as much as the humanists wanted to put man at the centre of the cosmos and shake off the medieval mental strictures, they were in some ways as devoted to received knowledge as their ancestors had been several centuries before. This put botanists like Mattioli in a quandary, since the ancients' categories were not always flexible enough to accommodate new additions. For example, maize was seen as simply a new kind of grain, and referred to as Indian grain, just as New World beans (belonging to the genus *Phaseolus*) looked similar to Old World beans (*Vicia faba*) and so were simply given that bean's name in most European languages. Other plants were harder to classify.

The classification system for plants and animals in early modern Europe was derived less from observation and more from cosmological beliefs. During the Middle Ages, philosophers elaborated a conception of nature as a reflection of the seeming 'natural order of things'. Mirroring hierarchy in human society, animals and vegetables were seen as having a place in what was called the Great Chain of Being. The 'lowest' foods, those grown underground or at ground level – carrots, onions, beans –were considered fit (and indeed, healthy) for the lower classes. Fruit, growing off the ground on trees, was more noble and hence fit for more refined people. Birds were especially prized, most particularly birds of prey, which flew high in the sky. In the *Decameron* (*c.* 1350), Giovanni Boccaccio

TVBERA rotundæ radices sunt, sine caule, sine folijs, flauescentes, vere fodiuntur. Cruda, & cocta eduntur.

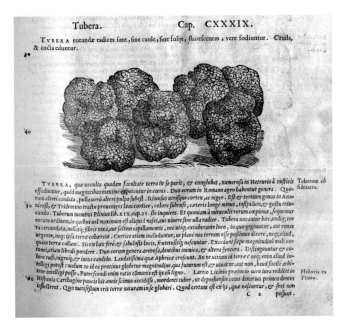

TVBERA, quæ occulta quadam facultate terræ in se parit, & conglobat, numerosa in Hetruriā & rusticis Tuberum cō effodiuntur, quòd magnatibus maximè expetuntur in cœnis. Duo eorum in Romano agro habentur genera. Quo-rum alteri candida, pulla verò alteri pulpa subest. Rimosus utrisque cortex, ac niger. Est & tertium genus in Ana-niensi, & Tridentino tractu prouenines laui cortice, colore subrubo, cæteris longè minus, insipidum, & gustu iniu-cundo. Tuberum meminit Plinius lib. x IX.cap.21. sic inquiens. Et quoniam à miraculis rerum cœpimus, sequemur eorum ordinem, in quibus uel maximum est aliquid nasci, aut uiuere sine ulla radice. Tubera nocentur hæc, undiq; ter-rā circundata, nullisq; fibris nixa, aut saltem capillamentis, nec utiq; exuberante loco, in quo gignuntur, aut rimas urgente; neq; ipsa terræ coherent. Cortice etiam includuntur, ut planè nec terram esse possimus dicere, neq; aliud, quàm terræ callum. Sic eis hæc ferè, & sabulosis locis, fruteto-sisq; nascuntur. Excedunt sæpe magnitudine mali co-tonei, etiam librali pondere. Duo eorum genera arenosa, dentibus inimica, & altera syncera. Distinguntur & co-lore rufo, nigroq; & intus candido. Laudatissima quæ Aphricæ crescunt. An ne uitium sit terræ s neq; enim aliud in-telligi potest: malum ne id ea protinus globetur magnitudine, qua futurum est, & uiuat ne aut non, haud facile arbi-tror intelligi posse. Putrescendi enim ratio cōmunis est ijs cū ligno. Lartio Licinio prætorio uiro iura reddēti in Hispania Carthagine paucis his annis scimus accidisse, mordenti tuber, ut deprehensis intus denarius primos dentes inflecteret. Quo manifestum erit terræ naturam in se globari. Quod certum est ex ijs, quæ nascuntur, & seri non possunt.

Tuberum cō sideratio.

Historia ex Plinio.

C 2

Pietro Mattioli's commentary of 1544 on Dioscorides' herbal includes this picture of truffles, with Latin commentary.

recounts the story of the nobleman Federigo whose lover, also noble, is coming to visit him. Having fallen on hard times he has little to offer her, and ultimately, desperate for a dish appropriate to her station, he kills and serves as lunch his prized hunting falcon.

This system of classification, like the Ptolemaic astrology of the day, was inadequate and contradictory; truffles are an excellent example. Growing underground in the manner of root vegetables, they should have been classified as a base ingredient, fit only for the 'vulgar' people, as the European elites called commoners. Yet the truffle's rarity – a result of the Europeans' inability to cultivate it – lent it some of what Peter Naccarato and Kathleen LeBesco call 'culinary capital'.

We can see this in a letter of November 1546 by Bernardo Machiavelli, son of the famous Florentine statesman Niccolò (a diplomat and head of the Republic of Florence's militia). When the Medici overthrew the short-lived republic in 1512, they removed Niccolò Machiavelli from office. He was later imprisoned and tortured, but was ultimately released and allowed to retire to his country home, where he wrote his political works, including his classic *The Prince*. His family shared in his shame, and years later one of his sons, Bernardo, attempted to ingratiate himself and his brothers with the Medici family (by then the dukes of Florence) by sending a package to Cosimo I de' Medici's personal secretary:

> With the present basket of straw with inside 50 pounds of truffles from Norcia which are very much well-kept, for Your Highness . . . and I would like to take the occasion to supplicate from my heart and spend four words to recommend to your Most Illustrious Patron the humble sons of Nicholò Machiavelli, who have been, are and always will be the servants of Your Lordship.

The attempts to shoehorn American plants into the old pre-Linnaean botanical system may have worked in favour of potatoes. The Spanish learned about the plant during their conquest of the Incas, and Alfonso Ceccarelli discussed it (without naming it) in a chapter entitled 'Whether [Truffles] Are Found in the New World'. Ceccarelli underlines the fact that while some commentators on New World foods have indicated that there are truffles there, these are not like European truffles but rather 'similar to mushrooms or chestnuts and are harvested from a plant similar to a poppy'. While some herbals referred to the plant as 'papa' (its original Quechuan, and Spanish, name), it is also 'tartuffali', the word for a certain

This letter of 1546 from Niccolò Machiavelli's son to the Grand Duke of Florence was sent with 50 lb of truffles from Norcia.

kind of truffle. The resemblance is a strong one, and indeed the name persisted long enough in Italian to be loaned to German; to this day the German word *Kartoffel* refers to potatoes. Mattioli notes that there are two types, white and black, and that they can be eaten cooked or fresh, but does little more than quote the ancients. In another sixteenth-century work on

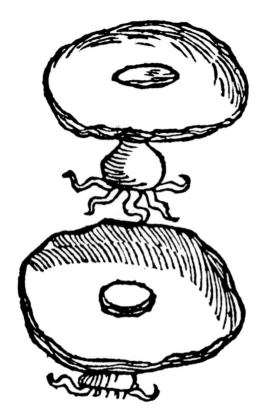

plants, Castore Durante's *Herbario Nuovo* of 1585, the author notes that 'this plant was to the ancients unknown.'

Durante's entry, to our modern ears, sounds downright odd:

Truffles

They generate black bile; when they are undercooked they damage the heart and nerves, and render urination difficult, generating as they do sand in the kidneys; they set in motion urinary problems, and in the same way are inimical to the teeth: and they generate a fatty liquid, paralysis, apoplexy and a frigid humour.

After this uninviting description, he gives the name of the truffle in Greek, Latin, Italian, Arabic, German, Spanish and French, and then notes that there are several species, differentiated by colour (both outside and inside). As to truffles' botanical category, Durante says that truffles 'are round roots (if one can call them roots) without foliage and without stem: but one should rather call them calluses of the earth'. Quoting Juvenal, Durante explains that truffles are 'generated' by the autumn rains, and by thunder. Truly bizarre, though, are his comments under the heading 'Qualities' (what we would call 'taste'):

> One cannot find in them any apparent quality. Those who use them in their food have a material that is disposed to receive all condiments which are given to them, as with every thing, that do not have in them any evident quality, and that are aqueous and bland.

Durante continues that truffles are made up of dirt and water and are 'without any taste. They generate melancholy and gross humours, more than any other food.' He does observe, though, that despite their qualities some people actually enjoy the tasteless roots: 'They are very much prized among the rich at dinners, as they believe that truffles eaten with pepper will excite the venereal appetites: cook under the ashes, wash, cook with oil in a frying pan, with pepper and orange juice.'

On Tubers and Wills

A postscript on Ceccarelli may serve as a metaphor for what has become rampant in the truffle market today: faking it. Ceccarelli, as we have seen, had been in the habit of making up quotations in his university days, but unfortunately he carried this over into his professional life. He was unsatisfied following his father as a small-town notary, and so, soon after graduating, he left for the city of Teramo in Abruzzo. It was perhaps there that he began what would be a long career in historical falsification. Ceccarelli's first jobs were for the minor nobility who needed documentary proof for their judicial claims to pieces of land. These were often hard to come by, so Ceccarelli, imitating the contorted Latin of the Middle Ages, made them up. He soon graduated to falsifying documents for small cities in boundary disputes with other municipalities, although he continued his private work as well, writing extensive (and equally invented) histories for princes who wanted a more glorious past for their families.

This line of work, while dangerous, paid well, and satisfied customers passed Ceccarelli on up the hierarchy. Through a cardinal in Orvieto, he made the acquaintance of Giovanni Maria Ciocchi del Monte, Pope Julius III. Julius took a liking to Ceccarelli and made him his personal doctor, even having him to stay in his palace on the Piazza Navona in Rome. It was in that city that Ceccarelli got to know other noble families, who appreciated not only his skill in 'research' but also his passion for astrology. By cultivating friendships with other important cardinals, he was eventually promoted to head astrologer at the papal court. History remained an important hobby, and his book *The Most Serene Nobility of the City of Rome* (1582) won him even more acclaim. While it was seemingly a history of the city's elite (larded with Latin quotations and imagined

Truffles in Norcia, an Umbrian city not far from Bevagna, where Alfonso Ceccarelli grew up.

noble predecessors for even self-made men), it was more of a publicity stunt to attract new clients in need of an invented past. Several months after it came out, Ceccarelli's father implored him to return to Bevagna; perhaps he knew that his son, while successful, was dancing near a precipice.

We have no record of Alfonso's response, but it is clear from his continued presence in Rome that he had no desire to return to the provinces. It was during that winter of 1582–3 that Ceccarelli mentioned to the Count of Anguillara that he had come into the possession of a 200-year-old will that proved that the Anguillaras should rightfully be not only counts of Anguillara, but also the dukes of nearby Ceri. The Anguillaras were overjoyed and paid Ceccarelli the princely sum of 3,000 *scudi* (25 *scudi* might have been the annual income of a schoolmaster in the late 1500s). After the Anguillaras presented the will in court, Ceccarelli went to the Duchess of Ceri and sold her another, apparently later will that revoked the former. The

45

sale of the document (for 6,000 *scudi*) was under way when somehow the faker was found out. He was arrested, imprisoned in the Tor di Nona in Rome, and probably tortured. Despite a lengthy defensive brief, protesting that he had acted only to protect the Church and its nobility, he was found guilty. His story ends with the following words from a chronicle of an order of monks in Rome:

> And thus at Hour VIII being day was celebrated and the said Alfonso received the Holy Sacrament. At Hour x he was conducted by the Ministers of Justice to the Bridge whence he was accompanied by our brothers, singing as is usual their litanies, and there his head was cut off. In the evening it was asked of our company of John the Baptist by his brother-in-law for his body, to bury it in San Celso, and he was permitted to do so.

Although the little hypogeous fungi had nothing to do with his execution, Ceccarelli's 'faking it' started early with his book on truffles. The French playwright Molière called his famous play *Tartuffe; or, The Impostor*, and as we shall see, fraud and truffles have a long history together.

3
Mycological Diplomacy

In 1614 Giacomo Castelvetro – a native of Modena, Italy, but having lived for years in self-imposed exile in England – copied out three versions of his essay *Brieve racconto di tutte le radici, di tutte le erbe et di tutti i frutti, che crudi o cotti in Italia si mangiano*. The name translates as 'A brief account of all the roots, of all the greens and of all the fruits, that raw or cooked one eats in Italy', although it is usually referred to as *The Fruits, Herbs and Vegetables of Italy*. This charming early modern account of Italian cuisine was half exhortation to the English to eat their vegetables, and half attempt at getting Castelvetro, then an old man, a pension from the Countess of Bedford so that he could live out his years without having to teach Italian to European nobility. Both attempts were failures, and Castelvetro's manuscripts languished for centuries, known only to scholars until a translation was published in 1989.

'This noble fruit'

Castelvetro's life had been anything but uneventful. As a strong-willed young man he had converted to Protestantism, following the example of his uncle Ludovico, whom he then had to follow into exile abroad, fleeing the Roman Inquisition. The two led a peripatetic existence, living for a time in Geneva but then moving frequently through France, Switzerland and Austria. The early modern food historian Gillian Riley notes that the older Castelvetro had a dyspeptic temperament that must have made his nephew attentive in procuring, if not preparing, delicious and innocuous vegetable dishes. Although Castelvetro's treatise on cooking was finished only 150 years after Platina's *On Respectable Pleasure and Good Health* came out, we can see more change in this period than in the five centuries preceding Platina.

Much has been said about the heavy use of spices in the Middle Ages and Renaissance, so it is important to note that Platina and the chef he copied, Maestro Martino, were already moving towards a more modern cuisine. Instead of being overwhelmed by spices, Platina's recipes were spiced with such 'flavour enhancers' as cinnamon, pepper and mace. As we have seen, if the recipes are burdened by anything, it is by the large dose of quotations from classical authors. It is hard to find a recipe among them without an anecdote from Cato and warnings about the food generating phlegm. Many sound strangely theatrical to our ears: 'artifice' was a goal for Renaissance cooks, and it is not unusual to find instructions on how to remove a pheasant's skin, cook the bird and sew it back up inside the skin. A medieval and Renaissance speciality was *blancmange*, a sort of white purée made from cooking chicken and pounding it into a pulp with almonds, milk, sugar and (of course) spices. Noble and royal tables were piled with food in quantities that could never be eaten, decorated with statues made from sugar and pies from which flew live birds. The theme throughout was elaborate spectacle and the concealment of what was natural.

This is why Castelvetro's recipes strike us as so modern, so surprisingly similar to contemporary cooking. Although he is often mentioned in the same breath as Platina, Castelvetro does not share the former's slavish devotion to citing the ancients and the shoehorning of good food into the nutritional theory of the day. Indeed, Castelvetro's essay has only a few references to food's effect on one's humours, and not a single classical citation. He demonstrates a remarkably modern empiricism. Talking about how some people think peaches are as unwholesome as they are delicious, he remarks:

> For this reason some steep them in good wine, which is supposed to draw out the harmful qualities, though I think myself that they do this more out of gluttony than because of any real danger. Peaches certainly taste much better with wine, and I notice that nobody ever throws away the wine that they have soaked in, or comes to any harm from drinking it.

The goal of the text is not to describe what is good for you, but simply what is good. Unlike many of his contemporaries, Castelvetro describes how commoners prepare the fruits of the field and garden without turning up his nose in class-induced disdain. The last entry in his treatise – and a relatively long one compared to others – is on truffles. Instead of Juvenal or Cicero, Castelvetro cites botanists, who 'tell us that this noble fruit is a kind of mushroom which grows hidden underground'. After giving a few lines from Petrarch's sonnet, he describes the use of a pig to find truffles. Here we note the difference from Platina's almost contrived account of a pig finding a truffle: instead of the farmer tickling the sow's ears and it magically backing off, Castelvetro tells us that the pig would very much like to devour the truffle it has found, but that the wily peasant, who has watched the pig closely, gives it a whack with his spade, 'and grabs the truffles for himself'.

Castelvetro also takes a moment to relate an amusing anecdote about a summer he had spent outside Basle. A young Swiss baron who had just returned from a journey through Italy asks Castelvetro about the curious custom some Italian gentlemen have of following pigs around in the woods, something he had apparently witnessed first-hand. Castelvetro attempts to explain that the pig is helping the gentleman (who is accompanied by a peasant to do the actual work) to find a treasure under the ground. 'What sort of treasure?' asks the baron. When Castelvetro replies *un tartufo* ('a truffle' in Italian), the baron, thinking Castelvetro has said *der Teufel* ('the Devil' in German), exclaims: 'Good heavens, how on earth can you bear to eat that monster?', although he calms down once Castelvetro has explained that it is actually a sort of mushroom.

A modern beverage that seems more like something that would have been drunk before the French Taste Revolution.

The French Taste Revolution

The radical cultural and culinary shift in what was considered good eating among European elites (poor people, of course, had always simply made do) between Platina and Castelvetro, in other words between the fifteenth and seventeenth centuries, is referred to by food historians as the 'French Taste Revolution'. A number of historical events set the stage for this dramatic shift – among them the invention of movable type, the discovery of the Americas, the Reformation, and empirical science disrupting the traditional association of cooking and dietetics – which made room in the kitchen for new foods.

Gutenberg's invention, according to the historian Elizabeth Eisenstein, was also the central cause of the Reformation. Luther himself referred with surprise to the speed with which his 99 theses, written in scholarly Latin, had been translated and spread via the printing press. Despite the Catholic Church's attempt to put the genie back into the bottle, the schism remained. Since the Middle Ages the Church had imposed limitations on Europe's eating, limits that were enforced by civil authorities with fines and punishment. These restrictions primarily concerned the difference between 'lean' and 'fat' days. Important feast days and the Lenten period required Catholics to give up certain foods, such as meat and fats that came from animals (primarily butter and lard). With the Reformation, European food customs started to diverge, and this divergence was driven forwards again by the printing press and the rise of national cuisines.

The Columbian Exchange was also a fundamental factor in the French Taste Revolution. What is strange is not the acceptance of such foods as tomatoes and potatoes – without which it is hard to imagine any European cuisine – but rather how uneven the speed of their incorporation was. Looking at Castelvetro's treatise, what he does *not* describe is fascinating. His final manuscript is from 1614, yet very few of the products of the Columbian Exchange appear in the text. While we do find a reference to the 'Turkish bean' (meaning 'bean from the Americas'; 'Turkish' at this time often indicated an exotic product, not necessarily from Turkey), we do not find the tomato, sweet or chilli peppers, potatoes or maize. We do know that Castelvetro was aware of this last product, though, because he listed it in another manuscript, a 'shopping list' of things to bring back from Italy for friends. These new products spread slowly and either took up places in recipes next to previously existing ingredients, or in some cases completely replaced them. Polenta, for instance, had previously been made with millet or buckwheat, but in the eighteenth century maize began encroaching on this recipe, and polenta became yellow.

Science, too, played a role. Previously the proper dietetics had been based on the idea of the Great Chain of Being and the system of Galenic humours. Although the differences between carbohydrates and proteins were still a mystery, books about cooking began in the late seventeenth century to disparage the 'old physicians' and their theories. The *Dictionnaire de Trévou* – an important eighteenth-century reference work – talks of acids, salts and tartars, not humours, when discussing theories of cooking. Science allowed, as the historian Jean-Louis Flandrin has written, 'the liberation of the gourmand'. No longer did the elite have to avoid products from below ground because they were not noble, or carefully balance fruits with spices.

Since ancient times, for example, sauces had been based on vinegar and had often been spiced, and there was no clear separation of sweet and savoury. These sauces gave way to condiments based on fats like butter, with a more restricted use of spices, except sugar (then considered a spice). The distinction between fruit and vegetables evolved along with the separation of sweet and savoury. Previously, all plant foods had been considered 'fruit' – indeed, most European languages have a linguistic remnant of these times in phrases like 'the fruits of the land' or 'the fruits of one's labours'. In the late sixteenth century 'fruit' came to mean sweet products. As the plantation system developed, first on the Atlantic islands and then later in the Caribbean and South America, sugar went from being an exotic medicine to an expensive spice to a widely used product. Sugar, which had in the Renaissance been used on pasta and in sauces, was pushed to the end of the meal, along with any sweetened products. Today it would be strange to find dessert anywhere but as the last dish, yet this is an inheritance of late sixteenth-century ideas about cooking.

Since Roman times the extraordinarily high prices of spices had made them more an object of distinction than a condiment. Integral to the cuisine of the elite and used in amounts that would nauseate the modern diner, they represented a kind of conspicuous consumption. Writing about spices in his book *Tastes of Paradise* (1992), the historian Wolfgang Schivelbusch gives us the spice allowances of the King of Scotland when he visited Richard I of England: two pounds of pepper and four pounds of cinnamon daily. As Europeans established trading centres and were able to acquire spices in ever greater quantities, the increased demand caused the prices to fall, thus allowing a larger part of the population to afford them. Whereas 'artifice' had been the keyword of Renaissance cooking, respecting ingredients' 'natural' qualities was the new vogue. One of the first cookbooks that clearly shows this new trend is the famous *Le Cuisinier françois* by François Pierre de la Varenne (1615–1678). 'La Varenne', as he is usually known, turned to herbs rather than spices, but also had a fondness for truffles, as is clear from their prevalence in his cookbooks.

This taste revolution had consequences other than prompting a new outlook on food and what constituted good cooking. The new vogue for 'natural' flavours and the dramatic decline in the price of spices and consequently in their value as a social distinguisher (something the anthropologist Sidney Mintz has called 'extensification') created an opportunity for the humble truffle. From a sort of gnarly vegetable – even Castelvetro suggests cooking them in the same way as potatoes in hot ashes, then peeling them and frying them in pieces – the hypogeous fungus was poised to become a condiment of the nobility. European elites needed to find something else to show their power in culinary terms, and the truffle presented itself as ideal. Unfortunately for most European monarchs, however, truffles were not evenly distributed on the Continent.

Savoy

Austria in the eighteenth century was not the small, neutral country of today, producer of ski champions and Mozart-kügeln; it was an enormous multi-ethnic empire that dominated central Europe. It was famous for having increased its territory through a careful policy of advantageous (and opportune) marriages. Indeed, a popular slogan went: 'Let others wage war; you, happy Austria, marry.' As will be clear from this section, this could have been paraphrased for another small, if increasingly influential, European power: 'Let others marry; you, happy Piedmont, send truffles.'

Piedmont was a territory that straddled the Alps where they met the Mediterranean – the name comes from the medieval Latin name for the region, Pedemontium, from *ad pedem montium*, 'at the feet of the mountains'. Piedmont was the

A pile of shavings of the *bianchetto* truffle (*Tuber borchii*), the great Alba white truffle's less-prized cousin.

domain of the dukes of Savoy. Their ancestors, like all European nobility, had probably been the most aggressive and retrograde inhabitants of the area, and their bellicosity meant they had taken control of the territory. Although small, Piedmont occupied an important position: it was at the intersection of the main north–south and east–west routes through Europe, and had neighbours that were both powerful (the French kingdom and the Austrian empire) and less powerful (various principalities on the northern Italian peninsula).

For centuries the dukes of Savoy had pursued the same short- and long-term objectives: enlarging the territory of the Duchy of Savoy, and moving up in the classifications of European states from duchy to kingdom. Carefully planned marriages were helpful, but war was the sure way to acquire new territory. The dukes of Piedmont were particularly active

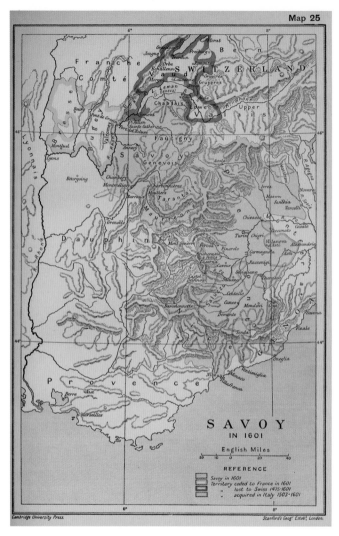

Map 25

SAVOY
IN 1601

English Miles

REFERENCE

Savoy in 1601
Territory ceded to France in 1601
„   lost to Swiss 1475-1601
„   acquired in Italy 1503-1601

Cambridge University Press.

Stanford's Geog.ˡ Estab.ᵗ London.

Map of the Savoy territory in 1601. The dukes of Savoy were slowly carving out a sphere of power at one of Europe's crossroads.

in the late seventeenth and early eighteenth centuries: when great European powers came looking for allies, successive dukes sent troops. Consistency was not the primary goal, but rather being on the winning side. Twice in these decades Piedmont fought with the Austrians, and twice against them (including one betrayal while the war was still in progress). Their territorial expansion came at the cost of Austrian losses, leading to a great deal of tension with Piedmont's much more powerful neighbour. The dukes of Savoy needed a way to placate and ingratiate themselves with the Austrians. Their tool: truffles.

The Dutch scholar Rengenier Rittersma, perhaps more than anyone else, has contributed to our knowledge of the social and political history of truffles, and his thorough and entertaining research gives us a glimpse of the truffle as an instrument of diplomacy. Rittersma has searched the Savoy archives and diplomatic correspondence and has found that as early as 1380 truffles were given as presents to important people in Piedmont (in this case from an ally to the wife of the Duke of Savoy). The practice of sending what we would today refer to as 'typical foods' to kings they wished to court continued into the eighteenth century. Frequently mentioned in the correspondence between the dukes and their representatives in Paris, Austria, Berlin and other European capitals were Piedmontese wines (unfortunately not specifically named), *rosolio* (a sweet liqueur whose aroma derives from different products, including orange, coffee and vanilla), jam from Mondovì, *fromage de Noël* (today known as Vacherin d'Abondance) and Piedmontese tobacco. What is surprising is that precisely in the period in which the Piedmontese dukes were adding the most to their domains – the late seventeenth and early eighteenth centuries – there are no references at all to truffles. As Rittersma is quick to point out, though, the importance of truffles in

European food culture has oscillated widely over the centuries. The ebb from 1670 to 1730 was about to turn into a flood.

In 1713 the dukes of Savoy had finally become kings (of Sicily), although in 1720 they traded up for Sardinia. In 1735 a preliminary peace was reached in the War of Polish Succession. Despite the name, almost none of the battles had taken place in Poland. Hostilities were principally between the French and Spanish Bourbon kings (with their ally, the Savoy king of Sardinia) and the Austrian Habsburgs. The Austrians were ultimately forced to make territorial concessions to the Savoys in the rich northern Italian duchy of Lombardy, and they were not happy about it, seeing the Savoys as upstarts and dangerous competitors in northern Italy.

It was with these diplomatic worries that Count Luigi Girolamo Malabaila di Canale arrived in Vienna in January 1737 as the Savoys' new ambassador. We can imagine him reaching the capital after a long, cold, bumpy carriage ride from Piedmont. After settling into his residence within Vienna's massive city walls (only much later were they torn down to make the beautiful Ringstrasse), he met his outgoing predecessor, Marquis Giuseppe Roberto Solaro di Breglio. We know from a fragment of a letter by Canale that the marquis taught him that food gifts were a good way to ingratiate himself with court society and get access to the highest levels of the Austrian government. But Marquis di Breglio's food of choice was the rock partridge, a bird related to the pheasant, which prefers to run away from predators than to attempt flight using its awkward little wings. Apparently it was Count Canale who hit on the idea of using truffles to gain favour.

Whether it was Canale's own initiative we cannot be sure, since the first request for truffles came from the Queen of Piedmont-Sardinia. She was a Habsburg, recently married off to the Savoy to try to bridge the diplomatic gap, and in 1738

The Savoys' secret weapon, the Piedmont white truffle (*Tuber magnatum*).

she sent several pounds of truffles to Canale, who was to give them to her brother, the Duke of Lorraine, in Vienna. Rittersma's careful searches in the archives of the House of Savoy reveal that from then the quantities increased, reaching 168 pounds of truffles in 1768, the year Canale died. Rittersma suggests that this was the result of a veritable European truffle mania in the eighteenth century. Indeed, he has found evidence

The Savoy sent out truffle-hunters and their dogs to monarchs across Europe.

that at least five European monarchs asked the Piedmontese for truffle dogs.

The first 'expedition' was in 1720: three dogs and an experienced truffle-hunter left for the court of the Prussian king. This was followed by a second group (this time four dogs and a hunter) sent to Paris. In 1751 one of the English king's sons, the Duke of Cumberland, wrote directly to Charles Emmanuel ii of Savoy and requested not only several dogs and a hunter, but also a man who knew how to train other dogs. What happened with all these expeditions is unfortunately not recorded in the archives. What we do know from these requests is that the truffle had gone from being a minor curiosity, boiled like many vegetables, to an object of desire in European capitals. Rittersma places this in the context of courtly society of the time: a rare gift gave prestige not only to its recipient in the eyes of the rest of the court, but also to the giver.

In this context of constantly shifting hierarchy, one-upmanship was of paramount importance. We know from Rittersma's research that the Empress of Austria, Maria Theresa, adored truffles. A letter from Canale's successor, Count Scarnafiggi, back to the court in Turin related that the empress had told him that the truffles were so delicious that everything else seemed insipid. Once the imperial couple was seen eating truffles frequently, the demand for them rose among those at court. Canale was well aware of this, as he wrote:

> It is some time that I spoke with you in my capacity of minister about the issue of truffles, and I will come to speak about it again, in order to tell you to send me once or twice a small quantity of truffles for distribution purposes. That would make it easier to get more familiar with some houses that it would be useful to have contact with. The most important ministers are grateful for this gift.

While Canale had, at the beginning of his term in Vienna, been instructed in gift-giving in a general way by

Late 19th-century etching (by Ludwig Friedrich, after Jacob van Campen?) of a man and a pig searching for truffles.

his predecessor, the practice was later codified and focused on truffles. Scarnafiggi and successive ministers of state were carefully instructed on whom to give truffles to, and how much. The great sums of money spent not only on their procurement in Piedmont, but also on their expedited shipping to Piedmont, show the seriousness with which the truffle gifts were taken. This is confirmed by the personal attention that the kings of Savoy gave to the truffle consignments. The white truffles of Piedmont were, as Rittersma comments, because of their rarity, 'simultaneously precious and as a gift object exclusively connected with the territory of Savoy'. He underlines that truffles were not just another gift, but rather part of coordinated diplomacy aimed at raising the House of Savoy's international profile. This worked, it seems, because after a few desultory wars (again against Austria), the Savoys became the monarchs of not only Sardinia and Piedmont, but also the whole Italian peninsula. This aggravated tension that already existed with their French neighbours; a rivalry that played out in the realm not only of politics, but also of truffles.

4
France and Gastrochauvinism

Marie-Antoine Carême had a tough childhood. A man who would become Europe's most famous chef probably started life without enough to eat. Details are sketchy, but it is possible that he had more than a dozen siblings. In the dangerous early days of the French Revolution, his father took him one morning to one of Paris's busy squares. He reportedly told his ten-year-old son: 'Nowadays you need only the spirit to make your fortune to make one, and you have that spirit. *Va petit!* – with what God has given you', and abandoned him there.

The boy who would later be so famous that he was known by his last name alone found work in a new kind of business, a restaurant. Despite the persistent myth that they were opened by the former cooks of all the beheaded nobility in Paris, the first restaurants were probably opened earlier, in the late 1770s. While they were originally places where people could have a restorative broth (hence the name 'restaurant'), these businesses soon began to offer a menu that customers could choose from. We take this for granted now, but it was quite the novelty in late eighteenth-century France. Although he started in a restaurant that served mainly meat, Carême soon graduated to a more prestigious establishment, a pastry shop. It was there that his career took off: people would come from

Tuber magnatum and *T. melanosporum*, the white Alba and black Périgord truffles.

all over Paris to see his elaborate creations made from spun sugar, puréed almonds and flour. In the days of plastic moulding it is hard to imagine the effect of Carême's food art; he could imitate famous sculptures or make models of Greek temples, not to be eaten but rather to be seen.

The theatrical element of cuisine was not a novelty: Renaissance banquets often featured music, dwarfs who told jokes and live birds flying out of pies. Carême, however, for the first time took the chef out of the kitchen and put him on a stage. Rachael Ray, Wolfgang Puck, Jamie Oliver and all other celebrity chefs and food personalities are descendants of this French orphan turned chef. Although Carême soon struck out on his own and began working for the movers and shakers of late eighteenth-century France, his work for the French diplomat Charles Maurice de Talleyrand-Périgord (whom we know today simply as Talleyrand) gave his career another large push forwards. Talleyrand was the son of a count from the Périgord region of southwest France; the area is popular with tourists today for the cave of Lascaux, but then it was one of France's underdeveloped areas. Its hilly terrain and thick forests made it poor at a time when agriculture was the most important occupation. One thing Périgord did have, though, was truffles.

A truffle-hunter's delight: a handful of truffles.

Vigo versus Carême

It is interesting to track the movement of the 'centre of gravity' of truffle culture over the centuries. In Roman times the most famous truffles came from north Africa. The writers of the Renaissance associated them with central Italy – Platina explicitly refers to truffle pigs in Norcia, a town high in the Apennines, near the city of Spoleto in Umbria. By the eighteenth century, the white truffles of Piedmont had gained a reputation as the premier truffle. As we have seen, while this is probably an effect of Piedmont's rising standing in European politics, it at least partly caused the dukes of Savoy to have a more important role in European diplomacy.

In 1776 a professor of rhetoric at the University of Turin, Giovanni Bernardo Vigo, published a book in Latin (*Tuber terrae: Carmen*, loosely translated as 'The Earth's Truffles'), modelled on Virgil's didactic poems about farming, the *Georgics*, of the first century BCE. It was the first salvo in what became a heated war caused by 'gastrochauvinism', a term the truffle historian Rengenier Rittersma uses to describe a certain country or region's claim to primacy, based on the production of certain foods or on culinary tradition. Like his work on the truffle in Piedmont's foreign policy, Rittersma's research on the gastronomic rivalry between France and Italy in the late eighteenth and early nineteenth centuries is fundamental to the history of the truffle.

Vigo was not a natural historian but rather a professor of rhetoric, which at the time meant that he (like the rest of the European elite) knew the Latin and Greek classics well. Vigo's decision to publish a book about truffles that explicitly refers to Virgil's *Georgics* makes it clear that his volume was not wholly didactic. Like Virgil, whose work was meant as a paean to the first Roman emperor, Augustus, Vigo

Périgord black truffles (*Tuber melanosporum*), always part of Carême's *pièce de résistance*.

was an unabashed patriot. Piedmont and its monarchs had been divinely appointed, something that was obvious from their rising status in the political realm, and also from their abundant natural endowments, including the white truffle: 'Because truffles are such particular fruits, and characteristic of our area of Subalpine Italy (even though not of all of it), whoever desires the best must search for them no further than here.'

In his dedication for the book, Vigo tells of his country-men who have gone to other countries that have truffles, but on returning home, declare that the white Piedmont truffle is certainly the best. France, then a culinary model for the rest of Europe, is explicitly singled out as having truffles inferior to those of Piedmont. Again Rittersma suggests that Vigo's statements are a reflection of political reality at the end of the eighteenth century: Piedmont's kings were realizing that their political future was located not west with France, but south on the Italian peninsula. Vigo was not alone in his patriotic grandstanding: in 1788 another scholar in Turin, Vittorio Pico, published a book in which he gave the Piedmontese white truffle the name *Tuber magnatum*, 'the royal truffle'. It seems that Pico proposed this name (which refers to the House of Savoy) at the 'suggestion' of the royal administrators. French naturalists objected strenuously to it, but it stuck.

Despite Vigo's book and Pico's somewhat pretentious name, primacy in truffle culture shifted to France at the begin-ning of the nineteenth century and remained there for more than 150 years. France's primacy was the result of three larger historical trends. The first is the French Taste Revolution, which began, as its name suggests, in France. The humoral system, and its prejudice against truffles' supposed generation of bile, eventually lost favour everywhere, but France was the first country where things could be eaten because they were

Typical countryside of the Périgord region. Truffles grew thickly in these oak forests.

good, and not necessarily because they were good for you. Another reason was the strengthened regionalism that was set in motion in France by the administrative reforms of the Revolution and, later, by Napoleon. Finally, members of the growing middle class were eager to show off their wealth and sophistication.

These wider historic developments had a huge impact on the French truffle market. Truffles, which had been banned for a time during the Revolution, became very popular in the Napoleonic years. Whether a sign of this new popularity, or one of the reasons for it, the famous French gourmand Jean Anthelme Brillat-Savarin (whom we know for his aphorism 'Tell me what you eat and I'll tell you who you are') called them 'the diamond of cuisine'. Napoleon's expedition to Egypt had sparked a rage for all things exotic, and truffles' fortunes rose accordingly; they were, as Rittersma says, 'the exotic product that comes from home'.

Whereas before there had been La Varenne's *Le Cuisinier françois*, now regional cuisines were codified and became popular, and regional cookbooks began to be published. The cuisine of the Périgord, as it became better known in the rest of France, introduced the French to the truffle. Périgord was definitely not the only French region to have black truffles (*Tuber melanosporum*), and they were probably not of the best quality in the country, but the region's general culinary renown built its reputation for truffles. The black winter truffle was now called the *truffe de Périgord*, even though – as demand for the truffles rose – many packages bearing this name were filled with truffles from other regions, or from central Italy. Just as it had become popular very quickly, so the Piedmontese white truffle saw its reputation fall. A French commentator in 1836 described it like this:

> In these truffles the taste of garlic very much dominates; it is far from the delicious aroma and from the aromatic, soothing smell with which true gourmet truffles are so richly endowed, this unpalatable and detestable truffle of Piedmont . . . There is no one gourmand who can delight in a taste which is so abominable . . . It is actually in France that the only real, good truffles are found, those who deserve the honour of enlightened gourmand, those which today have such a wide following.

Carême apparently agreed. One of his cookbooks, *The Royal Parisian Pastrycook and Confectioner* (published in English translation in 1834), has, as the first recipe in the section called 'Hot Pastry for the First Course', hot snipe pie with truffles:

> Pick and singe eight middle-sized snipes; take off their necks and feet, and then cut them in two. Take out the

backbone, and after wiping the inside with a napkin, place them on a sauter-plate [sauce pot], in which you have put four ounces of melted butter, the same quantity of grated bacon, a table-spoonful of parsley, two of mushrooms and four of truffles, all chopped very fine.

Truffles appear twice more in the recipe, sliced in between the snipes after they are placed inside the pastry, and then chopped, as a topping, after the pie is baked in the oven. The next recipe, for hot pheasant pie with truffles, is similar. Another, for hot quail pie with mushrooms, ends with the sentence 'To make this pie with truffles, follow the directions given in the two preceding receipts.' And so the section continues – hot pies with larks, à la Mongla, with beef palates, à la Financière, with poultry godiveau, hot fish pie, à la Mariner, à la Moderne, and so on, for pages – and almost the only ingredient the recipes have in common is truffles.

While it could simply be that Carême was caught up in the truffle mania of the nineteenth century, there is another possibility. Legend has it that when he applied for full-time employment at the home of the diplomat Talleyrand, he was asked to prepare a year's worth of extravagant menus, without ever repeating a dish. The brilliant young chef passed the test and joined Talleyrand's kitchen staff, where he learned many of the secrets that made him Europe's most famous chef later in his career. It is impossible to know when he had begun to use truffles as his most important ingredient, given the lack of documentation and Carême's own intentional mystification of his past. Yet perhaps it was in Talleyrand's kitchens that Carême developed his passion for the black truffle; as we have seen, Talleyrand's full title was Charles Maurice de Talleyrand-Périgord, and one cannot help but wonder if it might not have been the diplomat who introduced the chef to the underground treasure of his home region, and not the other way round.

When the Napoleonic wars finally ended and Talleyrand set off for Vienna in 1814 to represent France, Carême went with him. He was not present in the salons where the future of Europe was decided, but his contribution to French diplomacy was reputedly equally important. The chef prepared exquisite meals for the diplomats, and the story (perhaps started by Carême himself) goes that the concluding meal of the Congress of Vienna was one of his signature truffle dishes.

From Glass to Tins

In the hundred years between Carême's father abandoning him in Paris and France's production of truffles reaching one of its all-time highs, in 1898, truffles went from being a gourmet food that was the prerogative of the rich, to an everyday food that people at all levels of French (and indeed European) society enjoyed regularly. Two men were responsible for this astonishing democratization of Europe's shy mushrooms: Nicolas Appert and Auguste Rousseau. They were from opposite ends of France – Appert from near Champagne and Rousseau from Carpentras in the Vaucluse region – but they both worked to make truffles more available, to common people and to those who lived far from truffle-producing regions alike.

Nicolas Appert was born in 1750. His father was an innkeeper who dabbled in brewing and cooking. Young Nicolas worked alongside his father and developed his culinary skills well enough to become a chef for various members of the aristocracy. An entrepreneur before the word had been invented, he left a steady but boring job in the kitchen of a princess and started his own confectionery shop in Paris. An inveterate experimenter, Appert started a series of trials to make preserved food taste better and last longer. The principal methods of the day – salting or packing in vinegar or oil – all inevitably changed the flavour of the food they preserved. Truffles were the object of aristocratic, royal and imperial desire all over Europe in part because they were so hard to procure fresh: the delicate mushroom went bad quickly, and the only way for most Europeans to acquire them was, to paraphrase the sixteenth-century chef Cristoforo di Messisbugo, to have 'a fast steed or a full purse'. The availability of fresh truffles was limited to the areas within a six- or seven-day ride of the forest where they were found. Until Appert, that is.

Although he did not fully understand why his process worked, Appert discovered pasteurization decades before the publication of Louis Pasteur's *Études sur le vin* (1866). Appert would fill old wine bottles with broth or sauce, seal them well and put them in boiling water for several hours. What millions of people still do every summer – home preserving followed by processing in a water bath – seems like a relatively simple set of steps, but the odds were stacked against Appert's success. Given the lack of understanding of germs and bacteria, little importance was given to hygiene – indeed, Appert's devotion to empirical tests soon showed him that an almost maniacal insistence on cleanliness improved his results. He was bedevilled, though, by the closures of the bottles. His corks were too porous, but Appert realized this and had new ones made from layers of cork cut across the grain (and therefore with no openings for air to pass through) and glued together. Once in the bottles, the corks were secured with wire (much like, and perhaps inspired by, champagne bottles, with which he was surely familiar) and covered with pitch.

After three decades of experimentation, Appert's hard work paid off. He had perfected his process and was ready to present it to a governmental commission. The commission's members visited Appert's laboratory outside Paris – with its huge gardens and massive copper kettles, and hundreds of wide-mouthed glass jars – and watched him fill and process preserves. After a month, the commissioners returned to test the very same jars, and were surprised to find the contents as fresh as before. The French government offered Appert a large cash prize if he would publish his findings and explain the process. In the middle of the Napoleonic wars – the year was 1809 – the French were desperate for ways to increase their food self-sufficiency, cut off as they were by the British naval blockade. The French navy was soon using rations preserved using Appert's method, although there is some evidence that Appert himself was hardly patriotic, playing something of a double agent in order to secure an English patent. During a lull in the fighting he visited England, where he very probably sold his just-published process to a group of British investors. One of them had had experience with tin-plating, and soon the factory was turning out predecessors of the thousands of cans of preserved food.

Despite the explosion of the canning industry after Napoleon's defeat at Waterloo in 1815, Appert never enjoyed a happy retirement. His success came in waves, and unfortunately he died when he was in one of its troughs. In 1841 Appert breathed his last and was buried in the plain pine coffin that at the time was the sign of the burial of a destitute man.

Terroir and Thunderstorms

To understand the contribution of Auguste Rousseau to truffle history, we must first fast-forward to the Exposition Universelle of 1855 in Paris. While previous exhibitions of

All today's cans and tins are descendants of those invented by Appert.

the type (all precursors to the World's Fairs of later years) had featured products of culture and industry as their centrepieces (such as the Crystal Palace of Britain's Great Exhibition of 1851), France's agricultural products were in the spotlight in 1855. The official name – Exposition Universelle des produits de l'Agriculture, de l'Industrie et des Beaux-Arts de Paris – and the position of the word 'Agriculture' in that name made this clear. The Exposition was the brainchild of another Napoleon, Napoleon III, by this time the French emperor. Although we mainly remember it for the fact that the emperor requested the Bordeaux wine brokers to create a classification system for the region's wines (one that is still in use today), the Exposition was also meant to showcase all France's agricultural bounty.

The Exposition implicitly endorsed the growing regional loyalty that had been encouraged by the French Revolution, and with it the idea of *terroir*. This French idea, usually translated as 'the taste of place', holds that the unique combination of physical variables in any given location (soil type, precipitation, altitude) with cultural variables (traditional methods of

The first *truffière* ever to produce *Tuber melanosporum* outside its natural habitat in southern Europe. This *truffière* is near the town of Laytonville in Mendocino County, California. It began production in 1987.

food production) created food products that are unique to that place and not reproducible elsewhere. Champagne can come only from grapes grown in that region and vinified in the traditional way, just as Parmesan cheese can be made only in Parma and Newcastle brown ale only in Newcastle. Although this idea is actually quite difficult to prove empirically, it was very influential with gourmets in the mid-nineteenth century (and is still today). Such a concept is fine for products whose production can (within certain limits) be increased almost indefinitely (demand for Stilton cheese meant pasturing more cows in the fields around Stilton), it increased the demand for Périgord truffles, which were a relative rarity since they had to be found and gathered, not 'produced'. It was fortuitous, then, that at the Exposition of 1855 a young man from Carpentras (still today the centre of French truffle culture), Auguste Rousseau, presented an invention that would change truffle-eating for ever: the artificial truffle forest (*truffière*) and the tinned truffle.

Neither *truffières* nor truffles in a can were novel. As early as 1808 another Frenchman from Vaucluse, Joseph Talon, had discovered that transplanting the seedlings that had germinated under oaks in woods where truffles had been found would produce oaks that likewise had truffles around them. Talon, more of a businessman than a scientist, kept his secret to himself while busily buying up wasteland in Vaucluse and planting truffle acorns. He had no idea, of course, why it was exactly that this method worked; indeed, no one did except for a select group of scientists who understood that truffles were like any other mushroom. They began with spores, which grew into the fruiting body (what we call a truffle), which then developed more spores in its flesh. When ripe, these fruiting bodies gave off various perfumes to attract animals, which would eat them and spread the near-indestructible spores to another spot in the forest. Microbiology was still in its infancy, though, and very few people understood this process; the French referred to the mystery of truffle reproduction as *la grande mystique*. The mystery has been resolved now: the seedlings' roots had been inoculated with truffle spores. The process did not result in all the trees having truffles under them, but it worked well enough to be worth the time and effort.

As we have seen, opinions have varied greatly over the centuries as to what exactly the tubers are and how they are generated. The fourth-century BCE Greek philosopher Theophrastus remarked that 'there are people who believe that [truffles] are or can be raised from seed'. Two centuries later the Greek poet and physician Nicander wrote that he thought truffles might be 'silt modified by internal heat'. The Roman orator Cicero was slightly kinder in the first century BCE, suggesting that the desert truffles were the 'children of the earth', while the Roman naturalist Pliny the Elder just had questions: 'Whether this imperfection of the earth . . . grows, or whether

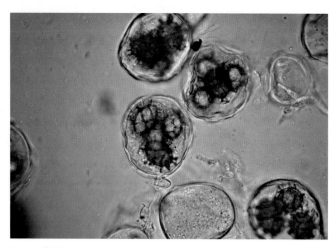

Spores inside their asci.

it lives or not, are questions which I think cannot be easily explained.' That thunderstorms contributed to the development of truffles was also suggested by the writer Athenaeus in the second century CE:

> Concerning these a singular fact is mentioned; it is said, namely, that they grow when the autumn rains come with severe thunderstorms; the more thundering there is, the more they grow, the presumption being that this is the more important cause.

Thirteen centuries later there had not been much progress in understanding the growth of truffles, with Hieronymus Bock's 'superfluous moisture' theory; while a contemporary of his, the botanist John Gerard, called truffles a 'tuberous excrescence'.

The first person to note the spores of truffles was the Italian scholar Giambattista della Porta (1535–1615), who wrote in 1588:

From fungi I have succeeded in collecting seed, very small and black, lying hidden in the oblong chambers or furrows extending from the stalk to the circumference, and chiefly from those which grow on stones, where, when falling, the seed is sown and sprouts with perennial fertility . . . in truffles, a black seed lies hidden.

Della Porta was ahead of his time, however. In 1623 the Swiss naturalist Gaspard Bauhin, echoing Bock, wrote that truffles were 'nothing but the superfluous humidity of soil, trees, rotten wood and other decaying substances'. The naturalist Sir Tancred Robinson, writing in about 1690, had a better opinion of the pseudo-tubers (he referred to them as 'a delicious and luxurious Piece of Dainty') but had clearly been reading Cicero and Pliny, since he says:

What these Trubs [tubers] are, neither the Ancients nor Moderns have clearly informed us; some will have them Callosities, or Warts, bred in the Earth: Others call them subterraneous Mushrooms.

Even as late as the mid-nineteenth century only one thing was sure about truffles, best expressed by Alexandre Dumas:

The most learned men have been questioned as to the nature of this tuber, and after two thousand years of argument and discussion their answer is the same as it was on the first day: we do not know. The truffles themselves have been interrogated, and have answered simply: eat us and praise the Lord.

It was in 1847 – around the time Dumas was writing about truffles – that Auguste Rousseau undertook a more systematic experiment, and his first truffles ripened under seven-year-old oak saplings, just in time for him to present his results at the Exposition of 1855. His prize, though, was not for creating one of the first *truffières*, but rather for a much-improved tin can for truffles, based of course on Appert's process. It was Rousseau's invention – which won him the Exposition's *médaille de première classe* – that allowed truffles to be canned on an industrial scale. But canning on an industrial scale required an industrial supply of truffles. This dramatic increase would come in the closing decades of the nineteenth century – not from the famed truffle region of Périgord, but from Provence.

The Men Who Planted Trees

Jean Giono was born in 1895 in the Provençal town of Manosque. An introspective boy, he loved to read, and despite having to leave school to work in a bank in order to help his family make ends meet, he continued to devour the classics. He was drafted into the army and fought in the First World War, returning to Manosque after the armistice to marry and settle down. His most famous work, *The Man Who Planted Trees* (1953), is the story of an unnamed protagonist who is walking alone through Provence in 1910. The protagonist is hiking through a desolate landscape, formerly home to villages, fields and babbling brooks but now denuded of trees and abandoned by all except a lone shepherd.

The shepherd introduces himself as Elzéard Bouffier, and invites the weary traveller to share his dinner and his cottage. After a meal with few words spoken, the traveller watches as the shepherd empties a sack full of acorns on to his table and, carefully examining each one for defects, counts out 100. The

Jean Giono's classic has sold more than a quarter of a million copies in English, and is often read on Earth Day in the United States.

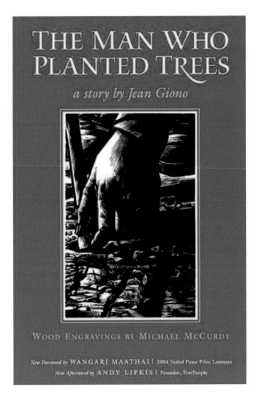

THE MAN WHO PLANTED TREES

a story by Jean Giono

WOOD ENGRAVINGS BY MICHAEL McCURDY

New Foreword by WANGARI MAATHAI | 2004 Nobel Peace Prize Laureate
New Afterword by ANDY LIPKIS | Founder, TreePeople

next day the traveller accompanies Bouffier and his flock, and notices the man sticking his staff into the ground and planting the acorns, one by one.

The traveller leaves, and the story jumps ahead to 1920. After fighting in the First World War, the protagonist has become discouraged with life and decides to take another hike through Provence. He comes to the same place, but this time sees thousands of small saplings. Bouffier is still there, but for fear that they will gnaw at his trees, he has got rid of his sheep and is now a bee-keeper. The protagonist is amazed to see streams running with water again, and grass growing

83

where before there was only scrub. Each year from then on the protagonist returns, and each year the forest is larger and stretches further. A government official takes notice of what he thinks is 'spontaneous reforestation', and part of the area is made into a national forest. People return and rebuild the old villages from their ruins, and Bouffier dies content, knowing that he has single-handedly brought a wasteland back to life.

Because Giono allowed the book to be printed by anyone who wanted to, its captivating story has enchanted thousands of readers. Many wondered if the story was based on real life, but Giono put this idea to rest in a letter to the mayor of a town near Manosque in 1957:

Dear sir,

Sorry to disappoint you, but Elzéard Bouffier is a fictional person. The goal was to make trees likeable, or more specifically, make planting trees likeable (this has always been one of my most fond ideas). And if I judge based on the results, it seems to have been attained through this imaginary person. The text which you read in *Trees and Life* has been translated into Danish, Finnish, Swedish, Norwegian, English, German, Russian, Czechoslovakian, Hungarian, Spanish, Italian, Yiddish and Polish.

I freely give away my rights, for all to publish. An American has come to me recently, to ask my permission to make 100,000 copies which he would distribute freely in America (which of course, I granted). The University of Zagreb has created a Yugoslavian translation. It is one of my works of which I am most proud. It does not bring me a cent, and this is why it is able to achieve the goal for which it was written.

I would like to meet with you, if that would be possible, to discuss practical uses of the work. I think it's time we created 'Tree Politics', though the word 'Political' seems very out of place.

Cordially,
Jean Giono

After Giono's death, his daughter described 'an old family story' as the basis of her father's tale. Until now, it had seemed as if that were true, but Rengenier Rittersma's research into the Vaucluse region in southeastern France has provided evidence for another story. We know now that *The Man Who Planted Trees* was at least inspired by a real man: his name was Durand Saint Amand, and he was from Vaucluse.

Vaucluse, while it has some low-lying areas perfect for agriculture, is dominated by hills, and its most prominent geographical feature is Mont Ventoux. The name alone – Mount Windy – gives those who have never visited an idea of the biting mistral that whistles down the Rhône valley and through the province. As well as having much terrain that was unsuitable for ploughs and a cold north wind to contend with (or perhaps precisely because of the latter), the Vauclusians over the centuries deforested the hillier parts of the province. The crushing economic depression at the end of the eighteenth century (the same one that ignited the French Revolution) accelerated the process, and the absence of a strong central government during the turbulent years of the Revolution only made the axes swing faster.

During the Napoleonic years stricter policies were enforced, but by then it was largely too late. One observer, describing Mont Ventoux and the surrounding province, spoke harshly to his fellow Vauclusians:

A now-reforested Mont Ventoux in the *département* of Vaucluse.

I would like to have the trumpet of the Last Judgment to mobilize the people of Vaucluse . . . : you who are so proud of your splendidly cultivated plains, go cross your desolate mountains; they are nearly all long since stripped of their decoration. Their slopes and plateaus have lost their cover of humus . . . The beds of the creeks and the rivers, once attractive and fertile valleys, are now devastated. Where you once saw vital woods and magnificent forests, there are today scrub and shrivelled trees left.

This was in 1866, just eleven years after the Exposition and Rousseau's discoveries of artificial truffle forests and an improved truffle tin. Yet only nine years later, more than 60,700 hectares (150,000 acres) – a sixth of the province's total surface area – had been planted with oak forests.

The man responsible for this incredible transformation was Durand Saint Amand. He was not a humble shepherd but rather the prefect (representative of the national government) of Vaucluse. He had already been trying for years to encourage the municipalities in his *département* to replant hilly areas with trees, but with little result. When he was informed in 1855 of Rousseau's discoveries, he realized how to kill two birds with one stone: reforestation with truffle-infected oak seedlings. He immediately sent an official letter to the mayors of all the towns in Vaucluse, telling them about Rousseau's *truffières*. But he also went one better, perhaps even overstepping his authority, and gave the mayors powers to use emergency federal funds to start this reforestation. Another factor made the reforestation campaign attractive: the phylloxera aphid had destroyed thousands of hectares of productive vines, leaving land that must be either replanted with vines at great expense, or abandoned. Saint Amand proposed a third solution: to replant with 'truffled' seedlings.

The results were as dramatic as they were immediate: in the blink of a historical eye – nine years – areas that had been desolate were covered with small trees. The presence of these special oak saplings had a cascade effect, giving homes to woodland creatures, creating a more favourable microclimate, improving water retention and preventing further erosion – just as in Giono's story. But the forests were not simply parks but rather part of the *département*'s agricultural production, since they were filled with truffles.

The total weight of truffles collected in 1875 – only twenty years after Saint Amand had sent his letter and released the emergency funds – was 450 tons, and by 1900 Vaucluse had reached a total annual production of 700 tons, an enormous amount given that total French production in recent years has averaged under 30 tons annually. Not only did these truffles feed eager gourmands in Provence, but also they were canned using the Appert-Rousseau method and shipped to the rest of France and the rest of the world. The last years of the nineteenth century saw a revolution not only in truffle

Holm oak (*Quercus ilex*) seedlings inoculated with *Tuber melanosporum* spores.

production (as more and more land all over southern France was planted with truffle acorns), but also in distribution.

The similarities of the two stories are striking: in Giono's, a shepherd plants acorns and in the space of a decade reforests a whole area in Provence. In the parallel historical version, a prefect helps mayors plant truffle-inoculated acorns and, in less than a decade, reforests a whole area in Provence. There is no evidence that Giono ever read Saint Amand's letter (he was born in 1895, forty years after it had been sent) or knew of the prefect's work. Surely, though, he had heard stories of the incredible forest that had sprung up almost overnight where there had been only desolation. Giono grew up in the *département* of Alpes-de-Haute-Provence, in the region of Provence, on the other side of Mont Ventoux from Vaucluse. Giono's boyhood was spent in Manosque, just over 80 km (50 miles) as the crow flies from the seat of Vaucluse's prefecture in Avignon. Whether *The Man Who Planted Trees* is purely Giono's invention or a pleasantly garbled version of the events of the 1860s, we can never know, but Vaucluse became and remains the centre of French truffle production – all because of the men who planted trees.

5
The Wandering Truffle

The most learned men have sought to ascertain the secret,
and fancied they discovered the seed. Their promises,
however, were vain, and no planting was ever followed
by a harvest. This perhaps is all right, for as one of the
great values of truffles is their dearness, perhaps they
would be less highly esteemed if they were cheaper.

Jean Anthelme Brillat-Savarin (1825)

Brillat-Savarin, widely recognized as the first modern gourmand, referred to truffles as 'the diamond of mushrooms'.
This is perhaps a more appropriate simile than he intended,
since it is well known that while there is actually an enormous
supply of diamonds, it is carefully regulated in order to keep
the prices of the not-so-rare gem artificially high. Anyone
wishing to find empirical evidence of the effect of the law of
supply and demand on prices need look no further than this
knobbly, perfumed fungus and the two examples it provides.
The first is the incredible late nineteenth-century increase in
truffle production and the effect on the price of the fungi; the
second is the crash in production between the beginning of
the First World War and the end of the Second, and its effect
on truffle prices.

As a result of the enormous rise in production brought by the artificial *truffières*, the price of truffles had fallen to 10 francs per kilogram by 1914, about the same price as potatoes that year, bringing them within the means of most French people, and available in quantity to the well-off. Indeed, when asked by a Parisian socialite how he liked his truffles, the French writer and gourmand Maurice Edmond Sailland (better known by his pen name, Curnonsky) replied: 'In great quantity, madam. In great quantity.' Prices dropped even further during the First World War, to 3 francs per kilogram. Ironically, many of the woods that had been replanted after extensive deforestation were now chopped down, since they were worth more as firewood than as truffle woods. The destruction wrought by the war was more than economic: a popular explanation for the decline in truffle harvest is that many truffle-hunters died in the trenches, and the knowledge of their secret groves and best spots for hunting died with them.

Much more likely than a drop in prices or the death of truffle-hunters as the cause of this precipitous drop in harvest

These summer black truffles are inexpensive, but their more prized cousins, the Périgord black truffle, can cost €1,000 a kilogram.

French truffle-hunter looking for truffles in one of Vaucluse's
many *truffières*.

was the long-term increase in urban population. All over
Europe, people from the country moved to small towns,
people from small towns moved to provincial capitals, and
people from provincial capitals moved to large cities. Indus-
trialization was at once the death of rural life and its gradual
embalming in museums dedicated to the peasantry, or in the
crystallization of France's restrictive system of geographical
indications for food, the AOC (*Appellation d'origine contrôlé*) system.
The French were leaving the countryside – and the under-
nourished, difficult life it represented – but wanted to preserve
its flavours to enjoy all year round, much in the manner of
Appert and his method of putting summer into a glass jar.

The truffle forests were either cut down or abandoned. It
might be hard to understand how an abandoned forest can

'decay', but truffles can grow only where there is not too much undergrowth to compete with their host trees. While France was still largely rural, peasants would cut brush for firewood and graze animals under the trees, thus keeping truffle harvests high. When the peasants moved to the city and took factory jobs, the forests became denser and darker, and truffles rarer and dearer. In 1868 some 1,534 tons of truffles were harvested in France; in 1920 that had fallen to 451 tons. In 2010 the total harvest was 32.3 tons, one-fiftieth of the harvest of 1868. As the availability of truffles dropped, their price rose correspondingly, so that in the 1950s the French poet Jean-Louis Vaudoyer could again comment on the price of truffles, opining that 'there are two types of people who eat truffles: those who think truffles are good because they are dear and those who know they are dear because they are good.' This decline continues today, although it is probably now linked to global climate change and uncontrolled harvesting.

Gold and Truffles

Truffles had also gone global – or rather, as has been more correctly said about the Americas, they were always there but only belatedly 'discovered' by Europeans. In the *Bulletin* of 1878 of the Torrey Botanical Society, just above an advertisement of the sale of a certain Mr James N. Bishop's herbarium (1,500 mounted plants, for which he thought he ought to receive four or five cents apiece), was a notice of a new truffle for North American flora. The *Bulletin* records the presentation at the society's recent meeting of this truffle, found on Staten Island, New York, by member W. R. Gerard. We know now that Mr Gerard was quite right in his statement that 'these fungi had always been regarded as rare in America: but that their rarity was probably more apparent than real, the plants not having been looked for.' The note goes on to describe how the truffle had been found in the town of Hugenot, on a sandy bank among the roots of alder trees. The anonymous bulletin-writer suggests that the truffle resembled the description of '*Tuber dryophilium*', but that it had deteriorated and so a watercolour of it was displayed instead.

Given the uncertainty of this first announcement, we should say that the first scientific description of a North American truffle was in 1899, when the Oregon white truffle (*Tuber gibbosum*) was 'discovered' by the amateur mycologist H. W. Harkness. Harkness was apparently quite the eccentric: born in Massachusetts and trained as a doctor, he had come west in 1849 to try to make his fortune panning for gold. He quickly realized that treating the gold-seekers was a more lucrative business, but he ultimately made more than $150 million (quite the sum even today) selling houses in the

Tuber gibbosum, the truffle discovered by H. W. Harkness.

Sacramento area. His money allowed him entry to California's elite society, and at the ceremonial completion of the Transcontinental Railroad in 1869 he carried the Golden Spike that was driven in to join the rails. Thereafter he 'retired' from his medical practice and devoted himself to rather disparate enterprises, including becoming superintendent of schools in Sacramento and writing about the age of the Lassen cinder cone in northern California. Another of these odd interests was California's hypogeous fungus, or truffles. Harkness's later years were occupied with this research, which he apparently carried out by bicycling around the countryside in a suit and top hat.

Published in 1899, two years before his death, Harkness's book *California Hypogeous Fungi* contains the first descriptions of some of the state's truffles, and constitutes the very first reliable description of the *Tuber* genus outside Europe. The noted mycologist Matt Trappe has commented on both the brevity and the inconsistencies in Harkness's description of what is now known as the Oregon white truffle. For example, Harkness wrote that he had discovered it 'under oaks', but this truffle forms symbiotic relationships only with Douglas firs; perhaps these were mixed in with the oaks that Harkness noted in his mycological journal. Although Harkness did not mention it (and it was formally described as a separate species only in 2010), there was another member of the *Tuber* genus on the West Coast as well. Confusingly, it is also often called the 'Oregon white truffle', since its Latin name is *Tuber oregonense*.

These were only the first truffles literally to be unearthed. In 1920 volume XII of the journal *Mycologia* carried the following announcement:

> In the last number of *Mycologia*, Miss Gilkey described two new American species of truffles, *Tuber canaliculatum* and

North American truffle, *Kalapuya brunnea.*

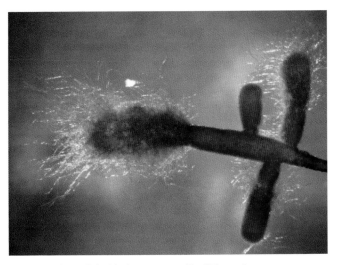

The mycorrhiza of the black summer truffle, *Tuber aestivum.*

T. unicolor, the latter based on material recently collected in the vicinity of New York City by the use of a dog trained in Europe. After working over this New York material, Miss Gilkey examined and carefully compared with it some specimens collected by Dr Shear in Maryland twenty years ago, and pronounced them similar but specifically distinct.

The article goes on to note that while '*Tuber unicolor*' had originally been described by Harkness, the description had been updated and the new truffle – yet another cousin of the European *Tuber* truffles – named *Tuber shearii*. The *Field Guide to North American Truffles* (2007) describes *T. shearii* as being 'palatable', while *T. canaliculatum* joins *T. gibbosum* and *T. oregonense* as 'delicious'. Despite their apparent gustatory value, the American truffles had little market value relative to their European counterparts, and they continue to be worth less than the better-known European truffles (despite the vigorous promotion of the Northwestern states of a potential goldmine in their soil). The challenges to European truffles in the market come from emigrants, the result of a green revolution.

A Happy Symbiosis

In 1880 the German botanist and biologist Albert Bernhard Frank was commissioned by the Kaiser to improve truffle cultivation, and in 1885 he published his research on the symbiotic association that he named *mycorrhiza* (from the ancient Greek words for 'fungus' and 'roots'). Frank observed that certain fungi that live in the soil form links with many species of plants. The fungus either wrap around or actually penetrate the plant's roots, and a mutually beneficial exchange occurs, whereby the fungus greatly extends the plant's root system, providing it with extra water and minerals, and the plant provides the fungus with a free meal of organic compounds like simple sugars. As research on mycorrhiza progressed, it was realized that in addition to truffle trees and legumes, most of the world's plants depended on such relationships with fungi to survive. After the collapse of the *truffières* in the interwar period, biologists built on Frank's work and investigated the mechanisms of associations that truffles form with trees.

During the 1960s a number of methods were developed to inoculate tree seedlings with truffle spores, and in 1977 the first truffles from a second-generation *truffière* were harvested in Aigremont, southern France. By the 1980s these techniques had been refined: the roots of the seedlings were sterilized and then coated with truffle 'inoculum', essentially ripe truffles that had been puréed and mixed with a simple syrup to improve adhesion. The seedlings were then planted in sterilized soil (to avoid competition from other, non-truffle mycorrhiza) and grown for a while in a sealed greenhouse until they were ready to be planted out in orderly rows in a prepared field. Today France's truffle production is still a fraction of what it used to be, but more than 80 per cent of the truffles produced are from these new *truffières*.

In the 1980s these techniques were further refined, and the inoculation of seedling roots moved from the laboratory to the commercial nursery. The first extra-European *truffière* – in northern California – produced a harvest of Périgord black truffles (*Tuber melanosporum*) in 1991. *Truffières* produced truffles in both North Carolina and New Zealand in 1993, and in 1996 in Taiwan. Demand in these areas grew with supply, thus encouraging even more *truffières*. According to Dr Ian Hall, a renowned scientist and creator of *truffières*, there are perhaps as many as 1,000 producing *truffières* outside Europe –

Greenhouse full of seedlings that have been inoculated with truffle spores.

in the United States, Australia, New Zealand, Chile, Israel, South Africa and Taiwan – almost all producing the Périgord black truffle. Yields are not as constant as those of the more established European *truffières*, being held back by climatic differences and a lack of trained truffle dogs.

'Without Mother Plant'

Despite the country's vast and ancient scientific literature, truffles do not seem to have been described in China until quite recently. They may have remained outside written accounts because they were restricted to rugged southwest China, and were a peasant food. The various Chinese names for the fungus – *wu niang tong* (fruiting body without mother plant) and *song mao fuling* (pine-needle fungus) among them – demonstrate the difference in cultural cachet that truffles have in China as opposed to, say, France. The first truffle in the *Tuber* genus (*T. taiyuanense*) was described in professional literature in 1985, although now more than 25 species have been found. In 1989 (the same year that a Périgord black *truffière* was planted in Taiwan) a small shipment of another black truffle – *T. indicum*, the most common Chinese black truffle – was sent to Germany for evaluation.

The Chinese truffle looks surprisingly like its French or Italian cousin: it has a rough exterior and a dark, almost purplish interior. Indeed, the resemblance to the Périgord truffle is so strong that it is often difficult even for experts to tell them apart; a microscope is needed to examine the 'spore ornamentation', not visible to the naked eye. The main difference is the price: the Chinese truffles are harvested using rakes, usually before they are fully ripe, and their aroma is much weaker than that of European truffles. *T. indicum* sells for $85 a pound wholesale, whereas the French Périgord truffle retails for up to $2,800. It does not take an experienced con artist to see the commercial possibilities, especially because the Chinese truffles easily absorb the strong aromas of European black truffles if the two are stored together for a few hours. Because truffle production in France and Italy (the main European producers, although Spain is now increasing its production) is very

The best-known Chinese truffle (*Tuber indicum*) for sale in a southwestern Chinese market.

low and demand worldwide is increasing rapidly, the temptation to mix Chinese truffles with the genuine (Périgord) article has been irresistible. According to the truffle enthusiast Gareth Renowden, some 'reliable witnesses report that a perfectly ripe Chinese truffle is the equal of its French cousin', but most truffles in China are harvested when unripe or deteriorate before they reach foreign markets.

There are several dangers, then. The first is that the truffle from China cannot express the *terroir* of a European truffle: it can never develop the same rich suite of aromas, simply because the pine forests of the mountains of southwest China are very different climatically from those of Mediterranean

Europe. Another danger is that honest labourers will be put out of business by cut-throat exporters of low-quality Chinese truffles harvested by exploited farmers. European truffle-hunters and *truffière* owners point out that they cannot compete with fraudulent truffles that are sold wholesale at a fraction of their goods' price. There is also a danger that the huge demand for Chinese truffles will damage an ecologically important – and fragile – part of China's ecosystem. South-western China is a 'hotspot' of biodiversity; the Chinese do not use pigs or dogs to find truffles, but rather simply rake, digging deeply into the ground under the pines with which the truffles associate. This not only disturbs the topsoil and creates erosion, but also severely damages truffle habitats, even destroying them. It is yet another ecological crisis created by a capitalist system without the normal societal-legal brakes in place.

Perhaps the most feared consequence of the importation of the Chinese truffle, however, is an unwanted biological invasion. Mycologists are not usually excitable, and the titles of their articles are relatively opaque to outsiders: 'Polymorphism at the Ribosomal DNA ITS and its Relation to Postglacial Re-colonization Routes of the Perigord Truffle *Tuber melanosporum*', for example, or 'Isolation and Characterization of Polymorphic Microsatellite Loci in White Truffle *Tuber magnatum*'. However, a letter in the journal *New Phytologist* in 2008 was quite different, almost hysterical relative to the normally restrained myco-logical tone: 'Is the Périgord Black Truffle Threatened by an Invasive Species? We Dreaded it and it has Happened!'

The letter defines invasive alien species as those 'introduced deliberately or unintentionally to areas outside their natural habitat' and which 'can cause a significant irreversible envir-onmental and socio-economic impact'. Although noting that humans have played a significant role in the movement of *Tuber* species in the last thirty years, and that truffles' habitats have

Bags of Chinese truffles that cost a tenth of the price of the Périgord black truffle.

never been static, the letter underlines that if faced with competition from *T. indicum*, *T. melanosporum* might be threatened with extinction. In addition to dismissing the Chinese truffle as organoleptically inferior, the writers describe it as 'dominant, competitive, and more aggressive' than the apparently more pacific Périgord truffle. The article concludes that

> *Tuber indicum* is a novel invasive species in Italy. More than a gourmet-market problem, it constitutes an ecological threat for the Périgord truffle, in particular because *T. indicum* can be found in China in ecosystems that are similar to those in central Italy . . . It is of vital importance to draw attention to *T. indicum* as an invasive alien species in Europe and to control its dissemination in *T. melanosporum* production areas.

What is striking about the passage above is its (unintentional) similarity to the racist discourse of late nineteenth-century Americans towards the Chinese: the xenophobic idea called the 'yellow peril' held that mass immigration of Chinese to the United States would overwhelm the native population and destroy a superior, but more fragile and less demographically vigorous people. The fear of the Chinese truffle is not restricted to the pages of dusty academic journals; a segment on the CBS show *60 Minutes* that aired on American television in 2012 drew popular outcry against Chinese truffles. The usual cast of corpulent French chefs and lithe truffle-hunters decried the invasive truffle: the former suggested that the mafia was behind the truffle's importation and that he would be killed if he revealed who cooked with it; the latter stated flatly that if *T. indicum* took over his *truffière*, he would be 'dead'.

Chinese truffle-hunters use rakes, not dogs, to find truffles and cause great damage to the fragile ecosystems where truffles are found.

What is ironic about the television programme's dire warnings and the mycologists' fear is that plants and animals have never respected the rather artificial boundaries drawn by humans, and humans have in the last six millennia (and especially in the last five centuries, as we saw with the Columbian exchange) been the prime movers in the dispersion of novel species. The term 'invasive' has a strongly negative connotation, but not much science to back that up. If we ignore the fact that no plant or animal has a static 'natural habitat' and use the definition as cited in the letter above, we would have to agree that vast portions of the United States have been the site of ecological devastation because of just such an invasive species. Forests have been cut down and irreplaceable topsoil eroded, all to produce the European invader *Triticum aestivum* (wheat). While it can be argued that this plant was intentionally introduced and grows only where it is planted, other invasive 'foreign' species (although of course species have no nationalities) include apple trees, earthworms and the beautiful wild flower known as Queen Anne's lace (the wild carrot, *Daucus carota*).

More ironic still is that while this letter obliquely acknowledges that 'humans have played an important role in the dissemination of some *Tuber* species', the writers seem much more concerned about the discovery of a single Chinese truffle in a Piedmontese *truffière* (and identified only through DNA analysis) than with the hundreds of artificial *truffières* that have been planted all over the world since the 1980s. The writers pose the question of how to 'control or eliminate *T. indicum* in Europe', and propose strict controls over any truffles arriving from abroad – and yet the same scientists are involved with publicly funded research to aid the development of *truffières* in other countries, with apparently no concerns about the impact of *T. melanosporum* on native ecosystems outside Europe.

Again, the sage commentary of Rengenier Rittersma:

> In this respect complex culture images are reflected: the Piedmontese truffle was a curiosity, representing a new political player (the kingdom of Piedmont-Sardinia in the eighteenth century), and it was supposed to accentuate that profile. The Périgord truffle symbolized the gastronomic hegemony of France, and the Chinese truffle has become, in a nutshell, the three-penny delight, as well as a botanical variant of the yellow peril.

Perhaps most ironic of all is the fact that China might be the ultimate source of much of the world's truffle biodiversity. The country's southwestern provinces, the mountains on its borders with Laos, Tibet and Vietnam, were protected from the last ice age and therefore provided refuge for flora and fauna that elsewhere perished with the encroachment of mile-high glaciers. Scientists have noted that this area is one of the main centres of origin of oaks, pines, beeches, birches, hazelnuts and willows. Chinese truffles have a wide range of host trees: the same oaks, willows and beeches that spread outwards from its southwestern mountains, but also pines, chestnuts and walnuts. Recent research has even shown that there are marked differences in flavour intensity according to where the truffles come from, effectively a Chinese *terroir*; and apparently *T. indicum* truffles that associate with broadleaved trees in the hot valleys smell and taste better than their coniferous mountain cousins.

After the last ice age, the trees (and animals, and fungus) that had survived in this area slowly spread out and repopulated Eurasia – 'invasive alien species' that thankfully were not mindful of invented notions of *terroir* or their supposed 'native' habitats. While the truffle family's genome and evolutionary

An officer in the Italian Carabinieri holds fraudulent truffles seized in an early morning sting.

family tree is still being untangled, it is clear that *T. indicum* cannot continue to be considered a foreigner of the 'European' truffles, but should be thought of as a cousin.

The Truffle Police

Fraud is still fraud, however, and in Italy, a country renowned for its high-quality food products, 'alimentary fraud' is a serious subject. The force charged with enforcing health codes is not a bureaucratic one but rather a division of the machine-gun-toting military police, the Carabinieri. The Carabinieri are usually known abroad for their Armani-designed dress uniforms, but they have an important role in Italy's armed forces (they are officially soldiers, not police) and crime prevention (their CSI units are famous in Europe). Since 1962 the Carabinieri division known as the NAS has been charged with protecting the safety – and the authenticity – of Italy's famous cheeses, wines and salamis . . . and of its truffles.

In May 2012, during routine checks in restaurants in Italy's gastronomic capital, Bologna, NAS agents found in a restaurant kitchen truffles that did not look quite like the truffle known in Italy as the *bianchetto* (*Tuber borchii*). While not as valuable as the famous Alba white truffle, the *bianchetto* still commands a retail price (depending on the harvest) of between €180 and €700 (about £140–550/$245–950) per kilo. The suspicions of the soldiers were later confirmed by morphological and chemical analysis at the University of Bologna, the results of which set in motion an investigation that had to uncover the source of the truffles before word got back that the product had been seized.

Despite being labelled *Tuber borchii*, the truffles were actually *T. oligospermum*, a species of whitish truffle that is similar in colour to the Alba truffle and the *bianchetto*, but easily distinguished on closer examination. The imposter has a very thin skin, so cutting into it reveals its identity immediately; cutting *T. oligospermum* into small pieces means that it is easier to pass it off as its nobler cousin.

The Carabinieri discovered that the restaurant in question had bought its truffle sauce from a producer in the city. An early-morning raid caught the producers red-handed: on one assembly line they were producing white truffle sauce with the real thing, while on the other they were using *T. oligospermum*. A block on mobile phone use, a quick search of records and quickly coordinated raids in other Italian cities revealed a much larger distribution chain than the Carabinieri had suspected. An importer in Pistoia, Tuscany, smuggled the valueless truffles in from north Africa (*Tuber oligospermum*, like *T. indicum*, is illegal in Italy because of the fear of fraud), then sold the contraband fungus wholesale to three different processors. These processors would purée the truffles, adding oil and synthetic chemicals that mimic the natural perfume of the Alba white truffle. They then sold these truffle products not only in Italy but also abroad (some jars were already labelled for sale in Brazil at the time of the raid). Captain Sabato Simonetti of NAS Bologna said that his investigators had seized more than 300 kg (660 lb) of the false white truffle purée, as well as warehoused raw materials worth more than €700,000 (£550,000/$950,000).

Ironically, it is legal to import certain truffles that it is illegal to sell in Italy. The Urbani Group, still run by the charismatic descendants of its founder, Paolo Urbani, imported Chinese truffles until recently. The fungi were washed, sorted and puréed because, as the company's CEO Olga Urbani said, 'they had the consistency of a cork.' Although illegal to sell in Italy, these Chinese truffles could be processed, packaged and exported. In other words, exactly what many truffle-hunters had feared – that Chinese truffles were supplanting their Périgord diamonds – was happening on an industrial scale. The Urbani Group sold Chinese truffles mainly to Germany, where companies that made pre-packaged meals wanted to give their products an inexpensive touch of class. Unlike the truffle bandits, the Urbani Group labelled their product clearly as containing Chinese truffles. It turns out that honesty did not pay; the market was never large for legal Chinese truffles, and because of lack of demand and other organizational problems, the company has given up on the Asiatic *Tuber*.

The Urbani Group is an excellent case study of how the truffle business, unlike any other agro-alimentary sector, still

Factory workers hand-washing truffles for the Urbani company, early 1900s.

has one foot firmly planted in the past. In the mid-nineteenth century Paolo Urbani began buying truffles from the many amateur hunters in the hills around the town of Scheggino, a little village nestling in the Apennines. The village straddles the River Nera, whose clear, extremely cold waters were the last place the truffle-hunters stopped (to wash their finds) before coming to Urbani's truffle-processing room ('not even big enough to call a "shop"', said Olga). As the demand for French truffles grew, many Italian *Tuber melanosporum* were shipped to Périgord and repackaged, labelled 'Produit de France'. Urbani realized that he needed to become a producer, not simply a wholesaler, and so he went to France, learned the Appert-Rousseau process and improved on it. His sealed glass containers, resembling something from a science laboratory, gave his truffles a longer shelf life and put Urbani (and Scheggino) on the map.

The next head of the company, Carlo Urbani, realized that America was the company's future. He ordered his cousin to go to New York to start a u.s. branch; the archives are full of plaintive letters from Paolo (later 'Paul'), insisting that the idea was hopeless because the Americans had no idea what truffles were, and begging to be allowed to come home to Italy. Carlo held firm, telling his brother to improve at explaining the value of the homely underground mushroom. Paul stayed and was ultimately able to convince Americans that the truffle was not a potato gone bad but a gustatory delight. He not only became a very rich man, but also was married four times, a testament to the amorous (however indirect) power of the truffle business.

The internationalization of the company proceeded apace, and Urbani now dominates the truffle market, processing more then 70 per cent of the world's truffles in its super-modern plant in Scheggino. Inside, workers clad in white carefully inspect the newly arrived truffles, and huge autoclaves prepare them for canning. And yet, despite the steel and glass, the Urbanis still depend on the grizzled men (and some women) who get up early in the morning and hike through the woods with their dogs, hoping to find a present from the earth. The Urbani group supports 200 families directly, but it buys truffles from more than 14,000 truffle-hunters in the area.

Carlo Urbani looked west for the future of the truffle business; his granddaughter Olga is looking east. The Urbanis have recently started selling their truffle products in China, but the company is looking for ever larger distribution to the country's luxury hotel chains, expensive restaurants and airlines. But while the Urbanis are looking forwards, they have not forgotten to look back. The company recently opened a museum dedicated to truffles and to the family business. The visitor can see letters not only from Paul – who threatens to pack his bags and take the first boat back – but also from U.S. presidents who received trufflegrams from the Urbanis, pictures of nineteenth-century truffle-hunters and artefacts from the company's past. 'Our name depends on our past', said Olga. 'What would we be without that?'

6

The Future of Fungi

Although scientists have broadened the definition of truffles to include a number of other genera, not just the European *Tuber* – *Leucangium* (some North American truffles), as well as *Terfezia* and *Tirmania* (the desert truffles) – all truffles have some things in common. They are just as much members of the Fungus kingdom as mushrooms, but unlike their vainer relatives, they are entirely subterranean until unearthed. All truffles are roughly circular in shape, although (as with other organisms) their final shape is influenced by their environment, in this case by the ground they grow in.

The ripening of the spores is for truffle-hunters the most exciting moment of the fungus's life cycle. Truffles, like all organisms, need to reproduce and, if possible, spread throughout their environment. The truffles' immobility is their Achilles heel, but they have developed a way to hitch a ride: their perfume. Truffles contain a complicated cocktail of volatile organic chemicals that combine to produce an irresistible scent. In addition to humans, many animals are enchanted by this heady odour, and those with better olfaction will dig up the truffle and eat it. We usually associate pigs with truffle-hunting, but dogs are much more common today. It is much easier to throw a dog in the back of a car than it is a pig, and the latter

A hillside being reforested with truffle-inoculated seedlings, a test plot of the University of Perugia in central Italy.

tends to want to eat the truffle, whereas a dog is usually content to receive a treat in exchange for the fungus.

When truffle-hunters get together and talk shop, there is much debate about what makes the perfect truffle dog. In recent years the poodle-like lagotto romagnolo has often been cited as the ideal dog by magazine articles on truffles. Truffle expert Ian Hall comments, though, that it is the lagotto's coat that gives it an advantage: 'That dog is perfect for early mornings in the Apennines, when it's quite cold, but in New Zealand a truffle dog doesn't need all that fur.' Hall tells about his brother's truffle dog, which went truffle-hunting in the *truffière* all by itself one night and left 5 kg (11 lb) of Périgord black truffles on the doorstep: 'Working the night shift: now that's a smart dog!' Ironically, the dog's first job had been in an airport, where as a sniffer he had not made the grade.

Matteo Bartolini, a truffle-hunter from the Umbria region of Italy, agrees that there is no one truffle dog breed. His dogs are of various breeds; the best one, Sole, is a half

Napa Valley
truffle-hunter
Bill Collins
and his lagotto
romagnolo, Rico.

bracco-pointer, half springer mix. 'What's important is not the breed, it's that the dog is smart, wants to please you and has fun going for walks looking for truffles', he maintains. 'Those are the right ingredients for a good truffle dog, not the breed.' Sole – who likes truffle-hunting so much that he occasionally eats one – is also called 'il Professore' by Bartolini, because he helps to teach the other truffle dogs. The professor's principal student is Zoe, a cute white-and-beige cocker spaniel mix who usually tags along with groups but is often looking more for pats than for truffles. 'She's *una vanitosa femmina da tartufo*', said Bartolini, 'a vain female truffle dog.'

When an animal eats a truffle, the spores pass through its body unharmed and 'germinate' after being deposited in the animal's faeces. The germinated spores produce thin cellular tentacles called mycelium, which snake through the soil looking for the root tips of their future tree hosts. Once the mycorrhizal relationship is established, the truffle fruiting body starts to develop, sending out a huge network of hyphae (like tiny roots, but even smaller than the smallest plant root hairs). These hyphae are able to reach into the smallest crevices between molecules that make up the soil, and draw water and minerals for the tree.

To understand this relationship better, think of an athlete who depletes his or her electrolytes on a long run. To replenish them, she drinks a tall glass of an energy drink, with a concentrated solution of these electrolytes. Imagine, though, that the glass were only half full of Gatorade, and that the rest was water: the athlete would have to drink two full glasses to get the same electrolytes. Imagine that each glass were simply water with a capful of energy drink: now the runner has to drink scores of glasses just to recover her electrolytes.

The process is similar with trees and other plants: in order to grow, they need minerals, such as calcium and phosphorus, that are dissolved in the groundwater. These elements are incredibly dilute, and so the tree is constantly absorbing water from its roots, drawing it up into the aerial parts and transpiring it through its leaves. It takes an enormous amount of water to gain the elements needed for growth: a mature oak tree can transpire 375 litres (100 gallons) of water a day. The truffle makes an agreement with the oak tree: you help me to eat, and I'll help you to drink. The fungus's extensive network of hyphae reaches into crevices that are inaccessible to the trees' roots, allowing it to drink enough of the requisite dilute elements; in return, the tree is happy to provide the truffle with sugars and other organic compounds that it needs to grow.

Truffles also render another valuable service to their host plants: they provide the muscle of the host's defence. While we often think of trees getting their nourishment from roots deep in the soil, many of the organic compounds are in the top 30 cm (1 ft) or so of the earth (the reason why truffles, too, are found near the surface). This puts the mighty oak in direct competition with the much smaller but still competitive annual plants that grow underneath it. It is a David and Goliath situation, but this time David gets whacked – by truffles. Among the volatile (easily becoming gaseous) chemicals that truffles emit are a large group of small but tough phytotoxic molecules, and these gases have been shown to cause leaf bleaching

The symbiosis between truffles and their host trees.

and root inhibition in the common European plant mouse-ear cress (*Arabidopsis thaliana*). The results are not definitive, but they suggest that the characteristic absence of herbaceous vegetation around truffle-hosting trees (referred to as *brûlé*, or 'burned') is not solely caused by truffles appropriating available nutrients for their hosts, but also because they are conducting chemical warfare against potential competitors. Researchers (mainly interested in a possible new herbicide) have not yet been able to isolate the chemicals the truffles may use to make the *brûlé*.

That said, Ian Hall cautions against viewing the relationship between truffle and host too rosily. A more accurate picture might include the truffle forcing itself on the tree as a nutrient partner. The exchange is perhaps not so amicable, either, with the truffle providing the bare minimum to the tree and being held back from further microbiological exploitation only by the tree's defences.

The Love Mushroom

Since ancient times, two truffle legends have been repeated time and time again, although without scientific evaluation: the first is that truffles are aphrodisiacs, and the second is that violent thunderstorms and lightning make the harvest better, somehow inducing truffle growth.

Galen, whom we met in chapter One, was not particularly interested in where truffles came from but noted that 'the truffle is very nourishing, and causes general excitation, conducive to sexual pleasure.' The Renaissance-era writer Michele Savonarola, in his *Book of the Things that One Eats* (1450), was more prosaic, saying that the strange tubers were 'a dish of old men with pretty wives'. The nineteenth-century French gourmand Jean Anthelme Brillat-Savarin was slightly more coy: 'Whoever says "truffle" utters a great word which arouses erotic and gastronomic memories among the skirted sex and memories gastronomic and erotic among the bearded sex'; and his countryman Alexandre Dumas held that truffles could, 'on certain occasions, make women more tender and men more lovable'. But it seems, however strangely, that these statements are false, and that centuries of truffle-induced passion have simply been caused by the placebo effect.

The idea that the truffle could have some sort of sensual effect on humans had been given some credence by the fact (popular in recent truffle writing) that one of its scent compounds, 5α-androst-16-en-3α-ol, is also found in the saliva of wild boar. The chemical, a steroid, was supposedly a pheromonal cue to wild sows, hence the diligence with which female pigs would search for truffles: they thought their Boar Prince was underground. By extension, suggest various articles in cooking magazines, this steroid might have some sort of chemical action on humans, making them more amorous. Unfortunately, science disagrees. A French chemist, Thierry Talou, found through a series of clever experiments that when given the choice between actual truffles, 5α-androst-16-en-3α-ol and a synthetic cocktail of the other major truffle aromatic compounds, pigs always ignored the 5α-androst-16-en-3α-ol and went for the more complete truffle scent, whether synthetic or real.

While truffles contain about 80 varied short-chain organic molecules (alcohols, aldehydes and ketones) that make up their smell and have near-unpronounceable names, there are some important ones, the presence (and decay) of which have a huge influence on the appeal of the truffle. The primary scent chemical in black truffles is CH_3SCH_3 (dimethyl sulphide). This and other sulphur compounds are the most important and the

A naturally occurring *brûlé* in Umbria.

most 'truffly', but they are also the most volatile; in other words, they are the first to become gases and drift away. European scientists recently spent five years decoding the genome of the Périgord black truffle, *Tuber melanosporum*, releasing their results in 2010. They found that the genome included the genes for making these volatile sulphur-containing metabolites. This means that the characteristic smell and flavour of truffles come from the truffle itself and not from soil microbes, as some had thought – the truffle gets the sulphur it needs from sulphates in the soil.

Another interesting recent discovery is that Chinese truffles, in addition to a large supporting cast of other volatile organic compounds, also contain dimethyl sulphide. Researchers reported in late 2012 that the compound made up almost 20 per cent of the total VOC content, something that came as a surprise to the truffle world, since it had previously been thought that *Tuber indicum* contained very little of this important chemical. This suggests again that the problem with these truffles is their ripeness, not an inherent inferiority in molecular endowment.

The chemistry writer Simon Cotton has pointed out that as these sulphur-based molecules start to dissipate, other molecules begin to dominate the smell of the truffle. One is 1-octen-3-ol, the main scent compound for mushrooms; indeed, humans perceive this compound as 'mushroomy'. That means that as truffles ripen (and decay slowly in refrigerators), they smell more and more like their humbler above-ground relatives. Dimethyl sulphide (above) is also a component of the smell of Italian white truffles, but the most important molecule there is a slightly bigger molecule called 2,4-dithiapentane (below), which resembles two dimethyl sulphide molecules 'joined at the hip'.

Particularly surprising is that dimethyl sulphide, while one of the most important compounds, must act in concert with all the other molecules to produce the smell of truffles. It is the Jimmy Page of the truffle aroma band, but it still has to be accompanied by Robert Plant, John Bonham and John Paul Jones to have that famous Led Zeppelin-esque truffle effect. Led Zeppelin analogies aside, in isolation, dimethyl sulphide can actually be an unpleasant smell. Cotton has pointed out that it is one of the main odour components of human flatus, is part of the bouquet of some beers and cheeses (including cheddar and camembert), is responsible for the smell of (over)cooked cabbage, and is produced by some sea sponges. Its aroma has also been associated with the charming scent of the dead horse arum lily (*Helicodiceros muscivorus*), an ornamental Mediterranean plant that attracts its primary pollinator (the blowfly) by replicating the stench of rotting meat. So much for romance and arousing 'erotic and gastronomic memories'. Luckily, when much diluted and acting in concert with other volatile compounds, dimethyl sulphide can have a pleasant effect on human olfaction.

The great eco-guru and scientist James Lovelock was the first to point out that the most important role of dimethyl sulphide in nature is to regulate the global climate cycle, as it is released from plankton in the oceans and degrades to sulphuric acid, the crucial precursor of cloud-forming aerosols – which means that on your next visit to the sea, you may smell dimethyl sulphide in the breeze. Incidentally, some sea animals – including albatrosses, penguins and seals – are attracted by dimethyl sulphide and use it to locate their food, although that is not known to include truffles. Perhaps simply inhaling the smell makes these animals 'more lovable'.

Discoveries by agricultural chemists have inadvertently created a lilting Frankenstein: truffle oil. Called the 'ketchup of the middle class' by the foodie Anthony Bourdain, this concoction is not, as one might expect, made by soaking truffles in oil, although that is indeed an easy way to make the oil, as the truffles' organic compounds infuse it. Truffle oil is rather a pale imitation: 2,4-dithiapentane is created in a laboratory and mixed into the oil in tiny quantities. The resulting oil has a flavour profile that is much stronger but also much narrower than natural truffle oil; white truffles, like their black cousins, have scores of chemicals that make up their aroma.

Pia Maria Mascaretti, *Cercando Tartufi* (Looking for Truffles), 2006, oil on canvas.

Another frequent claim about truffles, at least until recently, was that they were somehow generated by lightning and violent thunderstorms. We have already seen a number of quotations on the topic, but the Roman writer Plutarch's view is representative: 'Since, during storms, flames leap from the humid vapours and dark clouds emit deafening noises, is it surprising that lightning, when it strikes the ground, gives rise to truffles, which do not resemble plants?' Recent research suggests that this seemingly fantastic assertion might be true. Among the nutrients absorbed by truffle hyphae for their floral hosts are nitrates, nitrogenous compounds that are critical for plant growth (and the production of truffle spores), and which truffle hyphae can absorb much more efficiently than plant rootlets.

During thunderstorms, the incredible discharges of electrical energy that are lightning bolts are sufficient to break the bonds of atmospheric nitrogen and create other, more available nitrogen compounds such as nitrates. These associate with water molecules in the storm and fall to the earth, where they form salts with minerals in the soil, becoming natural fertilizers. It is likely that these nitrogenous compounds are much more common with frequent thunderstorms, and that the truffles in the area are effectively 'fertilized', however indirectly, by the lightning. Perhaps it is no coincidence, then, that the Bedouin tribes of the Negev desert in southern Israel call desert truffles 'the thunder fungus'.

Truffle Fairs, Truffle Festivals

The popularity of truffles has been both a cause and an effect of the creation in recent decades of fairs and festivals in Europe and across the world. The first festival was started in 1929 in the Piedmontese city of Alba. Before this date, this small city owed its fame to two rather dubious distinctions: it was the birthplace of Pertinax (who ruled the Roman Empire, briefly and viciously, for three months in 193 CE), and it had been sacked over the centuries by the Burgundians, the

Alfred Hitchcock, a recipient of the largest white truffle of the year, at the Alba Truffle Fair in Piedmont.

Lombards, the Savoy, the French and the Spanish. That all changed in 1929 when a hotel owner, the enterprising Giacomo Morra, decided to build (or, some would argue, invent) the city's reputation for truffle production. Alba had always been a minor centre for truffles, like any provincial city in France and Italy: truffle-hunters would bring their finds to a seasonal market and barter with wholesalers, who would then resell the truffles to larger processors. This kind of market can still be seen in Carpentras, one of the main centres of truffle production in France, in the famous *département* of Vaucluse.

Morra launched the Fiera d'Alba (Alba Fair, soon re-named the Alba Truffle Fair) in 1929. His careful and constant public-relations work made the fair a huge success rather quickly, and in 1933 the British newspaper *The Times* gave Alba the distinction of having the most exquisite truffles in the world, and Morra the title 'King of the Truffle'. Much as the clever De Beers advertising campaign of the 1930s had done, arranging for diamonds to appear in film scripts and on the fingers of newly engaged Hollywood starlets, Morra needed to place his truffles prominently. He began a tradition – which is still followed today – of sending the year's biggest truffle to someone famous. Rita Hayworth received the first trufflegram in 1949; subsequent recipients included Harry Truman (1951), Winston Churchill (1953), Joe DiMaggio and Marilyn Monroe (1954), Haile Selassie (1965) and both Dwight Eisenhower and Nikita Khrushchev in 1959.

Later years saw truffle packages sent express to Sophia Loren, Alfred Hitchcock, Pope John Paul II, Ronald Reagan, FIAT owner Gianni Agnelli, Mikhail Gorbachev and Luciano Pavarotti. While these later recipients had probably heard about the fungal 'Person of the Year' award, Harry Truman supposedly did not understand the gift. In a letter, almost certainly apocryphal, the American president is said to have written to Morra: 'Thank you for the potato. Unfortunately it must have gone bad during the journey and we had to throw it away.' There are, it must be noted, no truffles indigenous to Kansas.

All later fairs dedicated to the warty yet charming hypogeous mushroom are direct descendants of Morra's original fair. Some of the best in Italy are in Norcia, Città di Castello (where the Italian film beauty Monica Bellucci was once the festival's truffle queen, judging by photos of the event) and San Miniato. France's main fairs are in Carpentras, Sorges (in the heart of the Périgord region) and Trassanel (whose fair is called 'Truffles and *Terroir*'). These fairs have also spread to

In the future, climate change may necessitate more artificial *truffières* like this one, where the soil is adjusted with the addition of lime.

the New World, where the winning combination of truffles and wine is common. Indeed, one of the main American fairs takes place in the Napa valley, and features not imports from Europe but rather Périgord truffles from American *truffières*.

Two Truffle-hunters

When Ian Purkayastha was fourteen, he moved from Houston to rural Arkansas. With few friends and not a lot to do, he started foraging for wild edibles with his uncle. The Arkansas woods in which they hunted provided all sorts of gastronomic delights, but Ian's favourites were mushrooms. At a special dinner when he was fifteen, Purkayastha ordered black truffle ravioli with a foie gras sauce. He immediately fell in love with the taste of the earthy fungus, and wanted more. Fond of cooking, and eager to replicate the truffle ravioli, he began his search for truffles. His parents were unwilling to finance his truffle desires, so he searched Google until he found a purveyor willing to send him a pound of truffles direct. It was only after he had ordered the pound (0.5 kg) that he realized he might have gone a little overboard. He convinced his mother to drive him to the three fanciest restaurants in Fayetteville, Arkansas, managed – despite his young age – to sell all but a few of the truffles, and cooked with the remainder. 'That's where it all began', he says.

That summer Purkayastha ordered more truffles. His summer job was not washing cars or mowing lawns, but rather selling the fancy fungi to restaurants in Fayetteville, Houston, Austin and even Tulsa. He soon started both his own company, Tartufi Unlimited, and a culinary club at his school – Students for the Promotion of Exotic Culinary Experimentation – cooking huge meals for the other members, meals that featured ingredients like snails, cuttlefish, rattlesnake . . . and truffles.

The cold sell has been and continues to be Purkayastha's primary point of entry into the truffle market.

> When I first started out, my mother would drive me around to restaurants in Fayetteville and later Houston, and I would just walk through the back kitchen door, find the chef, and pull out my truffles. I had a wooden box that I would store the truffles in, and upon entering would give the chefs a gigantic truffle whiff, perfuming the entire kitchen.

He looked young but he was honest and forthright, had fresh truffles and was always available to deliver his product. At the age of sixteen, while touring colleges in New York, he was offered work with an Umbrian truffle company called PAQ.

'They were looking for a distributor and had seen my website,' Purkayastha explained, 'and soon after we met, they proposed we start immediately. I later realized that I wasn't supposed to *supply* their North American clients, I was supposed to *find* them, because they didn't have any at all.' Despite his age and inexperience, Purkayastha moved to New York (deferring college) to build PAQ's import business, peddling his perfumed wares to some of the city's most famous chefs. Eventually, after nearly four years with PAQ, Purkayastha wanted to go beyond truffles. PAQ was too focused on the fresh truffle business, and he wanted to expand his product offerings. He left the company and founded his own, Regalis Foods, to purvey not just truffles but also wild mushrooms, caviar and other speciality foods. While Purkayastha has expanded his range, truffles are still the core of the business. Some of his Périgord black truffles come from U.S. *truffières*, but most still arrive from Europe.

For most Americans, when thinking of entrepreneurship, there is nothing particularly amazing about someone who starts a business when young and later has their own company; what is amazing is that Purkayastha is now just nineteen years old. That first experience with truffles was when Ian was just fifteen. Since then he has gone from selling truffles in Arkansas to living in Manhattan's trendy West Village neighbourhood and being profiled in *Forbes*, which called him 'the prince of truffles'. Purkayastha is living proof that the future of the truffle business is not precariously balanced on the shoulders of the old men who roam the hills of Umbria and Périgord, but rather lies with a new, younger breed who realize the turmoil and potential fate of the industry in the face of global climate change.

On the other side of the world, Matteo Bartolini is another excellent case study for the evolution of truffle-hunting. He grew up just outside Città di Castello, a small Umbrian city near Perugia and right in the middle of one of the densest concentrations of naturally occurring truffle grounds in Italy. As a boy, Bartolini would go truffle-hunting with his father, although, 'I just wanted to break branches and throw rocks; I was more nuisance than help!' Bartolini's father hardly ever sold the truffles he found, but rather used them as something between gift and barter. He would give neighbours and friends truffles for their birthdays, or to express gratitude in addition to payment for services. This is still his habit, apparently: 'Just the other day I took him to the doctor and after his exam, he put a small black truffle in the doctor's hand', said Bartolini.

After studying economics (with a specialization in tourism) at university, Bartolini went to Italy's Adriatic Riviera, Rimini, and travelled all over Europe before coming back to Umbria with his fiancée in 2004. There was an opportunity to buy a small farm in the hills above Città di Castello, and he wanted to return to one of the most beautiful places in Italy. It was then that he rediscovered truffles; a friend and his family were visiting Bartolini's farm (which then produced only cereals), and, eager to have some time to chat alone, the two men rose early one morning and went truffle-hunting. 'He was entranced by it – how the dog worked with me, how we cleaned the truffles, and what we cooked with them – and I realized that even though I was young I was carrying centuries of tradition with me.' In 2007 Bartolini started a truffle school, something he considers as the link between Umbria's past and the future of a larger global village. The school has been a huge success, Bartolini has been recognized by the Italian Ministry of Agriculture, and his farm is the subject of research by the agriculture department of the University of Perugia.

The University of Perugia's test *truffière* at Matteo Bartolini's agriturismo, Ca' Solare, high above the Tiber valley near Città di Castello in Umbria.

Matteo Bartolini, and his dogs Sole (left) and Zoe, with a *bianchetto* truffle (*Tuber borchii*) that the latter had just found.

Despite the fact that Bartolini's farm is a small property, four different truffles occur naturally on its land, a fact that is highly unusual and goes against the received knowledge of truffle's eco-specificity, not to mention the idea of *terroir*. The University of Perugia has 'planted' one of Bartolini's former fields with 600 spore-inoculated oak seedlings; intercropped with the trees are electronic sensors, which take a reading of the soil's humidity, temperature and pH every hour. In a few years, when the seedlings have become saplings and Bartolini's head truffle dog, Sole, can start hunting there, the researchers will be able to get a better idea of the effects of these climatic variables on the production of truffles.

Ironically, Bartolini and Sole will not be the only ones hunting in the *truffière*: by Italian law, even though the property is Bartolini's, anyone can find truffles there. Truffles have always been considered a *dono di natura* (gift of nature), and as such – like wild game – they can be hunted even on private property, provided the hunter has a truffle-hunting licence from the Region of Umbria. One day, when Bartolini was taking students around from an American study abroad programme in Perugia, he crossed paths with two other truffle-hunters: 'Luckily, I have a better dog, and so I found the truffles they missed.'

Bartolini is working to change these outdated laws on private property, since he is also the President of the European Council of Young Farmers. 'It's not that I want to build fences around all the forests,' he explained, 'but we have to change our practices: careless truffle-hunters who roam all over other people's property don't cover up the holes that their dogs make. The truffle spores dry out and die and we have fewer and fewer truffles.' He assigns this decline in natural truffles not only to carelessness but also to the spectre of climate change. While the latter is hard to rein in, new Italian laws could limit poachers: instead of taking as many truffles as their dogs could find, they would have to make a deal with the landowner about best practice. 'But that is a long way off', said Bartolini. 'For right now, there are many more people looking for truffles than there are people actively managing *truffières*.'

Bartolini is hopeful about the future. In addition to better laws, he thinks research like that being done by the University of Perugia will help to halt the decline in truffle production. New laws will make people more sensitive not only to sustainable truffles, but also to sustainable agricultural and forestry practice. 'Learning about the truffle isn't just learning about a strange mushroom', he said. 'It's getting to know our past and pondering our future.' He thinks that while every truffle-hunter now has a little notebook in which to record his or her finds (the date, the approximate location, and so on), in the future this will be done with a GPS device. The hunter will simply hold the device over the hole the dog has dug, enter the type and approximate size of truffle, and click 'enter'. The following year they can simply return and find the exact spot, where truffles will probably have grown again.

But what about the famous 'Electronic Nose', already being tested in truffle research laboratories? These machines can be programmed to identify a precise mix of volatile chemicals, the 'signature' of a specific species of truffle; incredibly sensitive, they can even 'find' truffles under water. Bartolini, though, even if he had a combination GPS-tablet–Electronic Nose to use, would not leave Sole at home: 'Looking just at the machine, I would miss the nature. And Sole is not only my truffle dog, he's my friend – why would I want to walk through the woods every day without him?'

Truffle Identification

This short guide discusses the most common types of truffle. Note that 'peridium' refers to the truffle's 'skin', while the 'gleba' is its interior or 'flesh'.

Terfezia arenaria: 'Desert truffles', also called *Terfezia leonis*. Found in north Africa and southern Europe (especially Spain), they measure 3–10 cm (1¼–4 in.) across. Their peridium is yellowish to brown, and thin; the succulent gleba is whitish; and the aroma is very subtle. These truffles are eaten like a vegetable.

Tuber melanosporum: 'Périgord truffle' or 'black truffle of Norcia', the most highly prized black truffle. They grow wild and can be cultivated in *truffières*, but still can reach €1,000/kg (£790/$1,360). They can be as small as a pea but can reach the size of an apple. The warty black peridium can be ruddy at the base of the warts. The gleba is purplish-black and has white veins that turn pinkish when the truffle is cut. The aroma is intense and persists even if the flesh is heated, so the truffle can be cooked and not simply sliced fresh over food. It is ripe from early winter to early spring.

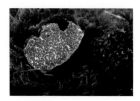

Tuber aestivum: Called the 'black summer truffle' or the 'Burgundy truffle', this is found in almost all European countries. An autumn variety is called *Tuber aestivum* var. *uncinatum*. Both resemble the Norcia black truffle, but with larger external warts and a beige–white interior. *T. aestivum* is very common from May to August. It is similar to the Norcia truffle in taste, although much less intense, and so less expensive. It ranges in size from 2 to 10 cm (¾–4 in.).

Tuber magnatum: The most highly prized truffle, now usually called the 'Alba white truffle', cannot be artificially produced. It is limited to certain regions of Italy and Croatia and so can cost more than €1,800/kg (£1,400/ $2,400). It has a smooth surface, usually cream to almost green. It has a thin peridium and a reddish-brown gleba, with white veins, and varies from walnut- to grapefruit-sized. Its complex aroma is destroyed by cooking, so it is usually shaved raw over cooked dishes.

Tuber borchii: Also referred to as *Tuber albidum*, this is another white truffle, often called *bianchetto* ('whitish') or *marzuolo* (literally 'March-ish') in Italian. It ripens between January and April and looks similar to the Alba white truffle, although its flavour is closer to that of garlic.

Tuber oregonense: Called the 'Oregon white truffle', this North American species has a roundish fruiting body and can grow to 5 cm (2 in.) in diameter. It is related to another North American truffle, *T. gibbosum*, associates with the Douglas fir, and grows on the west coast, from northern California to southern British Columbia. It has a whitish peridium and a white gleba that turns tan with age. It has a strong aroma, is widely used in gastronomy and can cost up to $220/kg (€161/£130).

Tuber indicum: Also referred to as *T. himalayense*, *T. pseudohimalayense* and *T. sinense*, although most mycologists believe these are all the same species. Very similar in appearance to the Norcia black truffle, it associates with a wide variety of broadleaved and coniferous trees, and is mainly found in China's southwestern mountains. Its aroma ranges from non-existent to subtle, and it is reputedly chewier than its European cousin.

Recipes

A Way to Prepare Truffles

We give the name Apicius to the author of a collection of recipes from the Roman Empire in the fourth or fifth century CE. The cuisine described in *De re coquinaria* is highly spiced and requires exotic ingredients that could be afforded only by the rich.

Scrape [brush] the truffles, parboil, sprinkle with salt, put several of them on a skewer, half fry them; then place them in a saucepan with oil, broth, reduced wine, wine, pepper and honey. When done retire [remove] the truffles, thicken the broth with roux, decorate the truffles nicely and serve.

Truffles in Ashes

The seventeenth-century writer Giacomo Castelvetro inserts this recipe into his account of truffle-hunting in Norcia, Italy. This is Gillian Riley's translation.

Truffles should be wrapped in damp paper and cooked in the ashes for about a quarter of an hour. Then peel them just as you would a baked apple or pear, cut them up very small, and finish cooking them in a pan with oil, salt and pepper. When they are nice and hot the truffles are ready to eat, and they are good to eat as they are, with just some lemon or bitter orange juice.

Black Truffle Soup

The chef Bartolomeo Scappi was private cook to two sixteenth-century popes. His cookbook *Opera* contains a number of recipes using truffles that were intended to be served towards the end of the meal, suggesting that he considered truffles to have digestive qualities.

Get truffles without sand, and let them stand under hot ashes for a half of a quarter of an hour, and then let them boil just a minute in wine and pepper. Remove the peels, cut them in pieces and put them in a glazed ceramic pot, or one of tinned copper, with much sweet oil, enough to stand covered with it, and with a little bit of salt, and pepper. Let them sauté slowly, and add then a bit of juice of sour oranges, or some verjuice, and cooked wine must, and let them finish cooking, and then prepare toasted bread, and sauté it in oil, having been dipped in a boiled composition of white wine, juice of sour oranges, sugar, cinnamon and cloves: after this composition has boiled, one puts the truffles with the broth in which they have cooked on top, and let them be served hot, but note that they should not be too salted, nor too insipid.

Hot Pheasant Pie, with Truffles

This recipe is taken from Carême's *The Royal Parisian Pastrycook and Confectioner*, first published in English in 1834.

Take two middle-sized pheasants, keep them for three or four days, in order to give them a high flavour, and then pick, singe and cut them up like pullets for *fricassee*; after which season them with sweet herbs, in the same manner I have mentioned before. When cold, garnish the bottom and the sides of the pie with four spoonsful of *godiveau* or fine force-meat, in which you have put two truffles chopped very fine: then put in the legs and rumps of the birds, and upon them four small truffles cut in two; next the fillets and breasts, and then a few more truffles; covering the whole with the seasonings before mentioned. Two laurel-leaves and some slices of larding bacon are then placed upon it; after which the pie is finished and baked for an hour and a half. The inside, after taking off the fat, is also covered with glazed demi-Spanish sauce, with truffles cut round like nutmegs. It must be served up the moment it is ready.

Insalata di Rossini

The composer Gioacchino Rossini was reputedly a gourmand from an early age, serving as an altar boy in order to drink wine. Rossini loved to add new elements to classic recipes, and called truffles 'the Mozart of mushrooms'. He also claimed that one of the three times in his life that he cried was when a truffled turkey, his intended lunch, fell off a boat into the water.

Take oil from Provence, English mustard, French vinegar, a little bit of lemon, salt and pepper. Crush and mix everything together, then add some thinly sliced black truffles. Truffles give this mixture a sort of halo, made to send me into the ecstasies of a glutton. The Pope's secretary of state, whom I met recently, gave me his apostolic benediction for this discovery.

Truffles Bologna Style, Raw, etc.

From Pellegrino Artusi's *Science in the Kitchen and the Art of Eating Well*, first published in 1891 and regarded as the first true Italian cookbook. Artusi compares the debate about black versus white truffles to the political fights in fourteenth-century Florence (between two parties called the Blacks and the Whites), then declares the black truffle to be 'the worst there is'. We can assume that he intends us to use the white truffle in these recipes.

After having dunked them in water and cleaned them off, as one usually does, with a little brush dipped in fresh water, the *Bolognesi* cut them up in very thin slices and layer them in a pan of tinned copper, alternating truffle slices with Parmesan cheese (sliced just as thin), starting however with truffles. Garnish with salt, pepper and a lot of good oil, and as soon as you've reached a good sizzle, squeeze a lemon over everything as you take it off the flame. Some people add a few pieces of butter; in that case, they don't put too much, which makes them too heavy. One can also eat fresh truffles sliced quite thinly and garnished with salt, pepper and lemon juice.

Crème of Desert Truffle

The film-maker, writer, photographer and truffle enthusiast John Feeney notes that for this recipe you will need not only a basket of white desert truffles, but also a female camel. If no camel is handy, use whole milk. This recipe is a modified version of one that Feeney tried when on assignment for Saudi Aramco World.

Put the 9 or 10 medium-sized white desert truffles in cold water for 10 minutes. Throw out the water and loose sand and cover them with water again. Repeat. Carefully scrub to remove all remaining. Peel, conserve peels. Cover peels with 2 cups (450 g) whole milk and simmer for 10 minutes. Cool, gently pour off the milk, leaving the sand behind. Conserve milk.

Roughly chop all but two of the peeled truffles. Put 1 small onion and 2 cloves of garlic (all peeled and roughly chopped) in 4 cups (900 ml) whole milk (not the milk reserved before). Bring to a boil, boil for 5 minutes and add the chopped truffles. Simmer gently for another three minutes, no longer. Purée the mixture and set it aside.

Make a white roux using 2 cups (450 ml) milk, 1 tbsp butter and 2 tbsp flour. Do not boil; when the roux has thickened, pour in the very hot milk, a little at a time. Stir continuously and simmer very gently for another 10 minutes.

Slowly stir in the puréed truffle mixture until it is absorbed into the sauce. Drop in 1 beef stock (bouillon) cube and ½ tsp sugar. Add salt, white pepper and ¼ tsp cayenne pepper. Gently, so as not to raise any remaining sand from the bottom, stir in the milk the skins were boiled in. Stir in ¾ cup (170 ml) single (light) cream. If the soup seems too thick, dilute with a little more milk.

At the very last moment before serving, so as to obtain the maximum truffle flavour, take the two peeled truffles you have set aside and grate them, using a rasp or the finest part of a kitchen grater, directly into the soup.

If you have been lucky enough to find one or two truffles with a pink interior, the crème will have a seductive pink tinge. It is especially good served with warm cheese-straw pastries.

Truffe à la Vaucluse

Courtesy of the Vaucluse tourist office, according to which this recipe is the 'simplest way to taste truffles'.

Toast four slices of country bread, cover them with 80 g truffles (sliced, not too thinly), then pour a trickle of olive oil over, and add a pinch of French sea salt from the Camargue (*fleur de sel*).
Serves 4

Truffled Butter, Oil, Eggs or Cheese

From the truffle researcher and hunter Matt Trappe.

Place cleaned and dried truffles in a paper-towel-lined, sealed plastic container in the refrigerator. For truffled butter, place sticks of butter in the container with them (with wrapping loosened). Do the same with cheese (mild cheeses showcase the flavour best). Raw or hard-boiled eggs use the same process; the aroma easily penetrates the eggshell. Truffled raw eggs retain much of their truffle flavour through cooking. For truffled olive oil, use a larger plastic container and place an open bowl of oil in it, with the truffles around the outside of the bowl. The key point with all these methods is that the truffles never actually contact the food, only the aromatics are absorbed. If you add truffles directly to raw foods, they must be consumed within a few days or they will get funky.

Truffle Puffs

Intended to be made with Oregon truffles, this recipe is from the truffle enthusiast Frank Evans, who said: 'This recipe came to me while my wife, Karan, and I were eating clam puffs at a restaurant on the Oregon Coast. I tried an experimental batch on some friends one Sunday afternoon and I have been pleased with it ever since. I have made a batch with a variety of truffles in different puffs, allowing comparison of the flavours.'

Lightly sauté 1 finely chopped shallot in a little butter. Add fresh spring onions (scallions), fresh parsley and 1 or more walnut-sized truffle (finely chopped) as you remove the pan from the heat. Mix this into 8 oz (225 g) softened cream cheese. Toast a piece of bread and crumble it finely, then work the crumbs into the cheese and truffle mixture. Set aside.

Separate a sheet of filo pastry (about eight layers' worth). Cut the sheet into 3 in. (7.5 cm) squares. Drop about a tablespoonful of the filling into the centre of each square. Fold the corners in to the centre to just cover the filling. Bake in a very hot oven (475–525°F/250–275°C) just long enough to brown the pastry and heat the filling (2–3 minutes).

Grated Truffle over Pasta

Matteo Bartolini ate this pasta as a boy and still makes it for guests after truffle-hunting.

Cook long, thin pasta like fettuccine in salted water. When the pasta is almost ready, gently sauté truffle shavings in butter for two minutes. Add a few spoonfuls of the pasta water, drain the pasta and mix with the truffle butter. Grate a few slices of truffle over the top of each serving.

Risotto with Truffle

A modified version of a classic truffle recipe from Piedmont, courtesy of the Capital and Canberra Region Truffle Festival.

900 ml chicken stock
40 g unsalted butter
125 g unsalted butter
1 small onion, finely chopped
300 g Carnaroli rice
175 ml white wine
25 g Parmesan, freshly grated

Heat the stock in a pan until simmering. Gently melt half the butter in a small saucepan, then take off the heat. Finely grate or shave most of the truffle (a Microplane is ideal or use a truffle shaver) and add to the butter. Mix well and reserve. Melt the rest of the butter in a heavy-based saucepan. Add the onion and cook gently until softened. Add the rice and cook, stirring with a wooden spoon over a medium heat until the butter fully coats the rice (approx. 3 minutes). Add the white wine and simmer until absorbed. Add the hot stock a ladleful at a time, stirring well between each ladle. Continue adding stock and stirring for approximately 20 minutes. When the rice is tender and moist and all the stock has been added, remove from the heat and allow to rest for 30 seconds. Add the remaining truffle butter and grated Parmesan, stirring vigorously and season with salt. Serve topped with shaved fresh truffle, about 2 g per portion.

Frisée Salad of Black Truffles

This recipe was a hit at the Napa Truffle Festival in 2013. It is the invention of Dominic Orsini, winery chef for Silver Oak and Twomey cellars.

2 eggs, beaten, plus 6 whole
1 cup (140 g) plain flour
2 cups (100 g) breadcrumbs
1 oz (25 g) fresh black truffles
1 tbsp extra virgin olive oil
2 tbsp shallots, finely chopped
1 lb (450 g) foraged mushrooms (yellow-foot, hedgehog and black trumpet; or substitute chanterelles, oyster mushrooms or shiitake)
2 tbsp chopped parsley
4 potatoes of each kind: red, purple, Yukon gold (or other waxy potatoes)
2 quarts (1.9 litres) canola or light vegetable oil
4 heads frisée lettuce
3 oz (85 g) truffle cheese
extra virgin olive oil, for drizzling
1 lemon
2 tbsp chives
salt and pepper

Fill a medium pan with water and bring to the boil. Place the whole eggs in a sieve that will fit into the pan, and lower the sieve into the boiling water. Cook the eggs for exactly 6 minutes. Remove the eggs from the pan and submerge in a bowl of iced water. Cool in the water for 10 minutes. Gently peel off the eggshells and lightly rinse the eggs, being careful to keep them intact. Coat each egg with flour, dip into the beaten egg and then coat with breadcrumbs. Put aside until ready to fry.

Peel the outer rough surface of the truffles with a vegetable peeler. Dice the peelings very finely and set aside to complete the salad at the end.

Heat a large frying pan over a medium-high heat. Once hot, add the olive oil, then the shallots. Give a quick stir, and once the shallots have toasted to a golden colour, add the mushrooms (cut or torn into bite-size pieces). Let them sauté together until all the moisture that is released from the mushrooms boils away. Be sure to stir frequently. This should take 3–5 minutes, depending on how wet the mushrooms are. Once ready, stir in the truffle peelings and the parsley. Set the mushrooms aside in a warm place.

Boil the potatoes until tender, drain and pat dry. Heat the canola oil in a high-sided pan to 350°F (180°C). Gently crush each of the potatoes, but leave them intact. Lower the potatoes into the hot oil and fry until crispy. Remove from the oil and place on a plate lined with paper towel. Season with salt and pepper.

Wait for the temperature of the oil to return to 350°F (180°C). Place the breaded eggs into the hot oil to fry until golden brown in colour, one or two at a time. Remove from the oil and place on the plate with the crispy potatoes. Season with salt and pepper.

Wash and trim the frisée lettuce of its dark green tips, then cut into bite-size pieces. Place the lettuce and truffle cheese (shaved or crumbled) in a salad bowl. Toss the salad with olive oil, salt, pepper and a light squeeze of lemon juice to taste.

To serve, place a small pile of sautéed mushrooms in the centre of each plate. Place the salad on top, then the crispy potatoes around the salad. Using a paring knife, make a small incision in the centre of the egg, and with your hands gently break it open. The yolk will begin to ooze out of the centre. Place the split egg on top of the salad and finish the dish with a sprinkling of chopped chives and shavings of black truffles on top. Serve immediately.
Serves 6

Black Truffle Ravioli with Foie Gras Cream Sauce

Ian Purkayastha's version of the recipe that turned him on to truffles.

3 large Yukon gold (or other waxy) potatoes
120 g butter (preferably truffle butter)
600 g wheat flour, oo flour if available, plus extra for dusting
5 eggs, plus one extra egg yolk
1 tbsp extra virgin olive oil
120 g fresh winter black truffle (*Tuber melanosporum*)
200 g foie gras torchon
1 cup (225 ml) double (heavy) cream
⅓ cup (75 ml) sweet wine (Tokaj Aszu works well)

For the ravioli filling, set 5 cups (1.2 litres) water to boil in a large pan. Add the potatoes and boil until tender. Drain and remove the skins, crush the potatoes until soft, then add the butter and mix thoroughly. Leave to cool.

To make the pasta, pulse the flour, 5 of the eggs, egg yolk and oil in a food processor until all thoroughly mixed. Put the mixture on a floured cutting board and roll the dough thinly into long sheets (run through a pasta-maker).

Shave 30 g of the truffle into circular medallions. Spoon the potato mixture in circular clumps 2 in. (5 cm) apart on half of the pasta sheets, and top each with 1 or 2 truffle shavings.

In a mixing bowl, thoroughly beat one egg yolk. Brush the beaten egg yolk over all exposed areas of the pasta sheet. Place an empty pasta sheet on top, matching up all sides evenly. With a cookie cutter, stamp out each ravioli individually.

In a large pan, set 10 cups (2.3 litres) water to boil. Add the ravioli to the boiling water and cook for 2 minutes. Drain the ravioli, reserving ½ cup (110 ml) of the pasta water, and put in a clean pan.

For the foie gras sauce, cut the foie gras torchon in half. Wrap one half in clingfilm (plastic wrap) and freeze. Put the remaining piece in a frying pan, add the cream and wine, and emulsify over a medium heat until thoroughly mixed. Spoon over the cooked ravioli.

Remove the other piece of foie gras from the freezer. Shave the remainder of the black truffle and the foie gras over the ravioli, and serve.

Serves 6

Sauce Périgourdine

A good basic sauce that can be used in a variety of dishes to provide truffly flavour. This version of a classic recipe is courtesy of Ian Hall.

Wash 1 kg fresh truffles well to remove any earth. Place the truffles in a large pan and add 250 ml port, 250 ml Madeira and 5 litres strong brown stock. Bring to a boil, reduce the heat and simmer very gently for 15 minutes. This will produce a well-flavoured truffle juice that can be used as a base for many recipes; to keep the truffles for later use, place them in preserving jars, cover with truffle juice, seal tight and sterilize in a hot-water bath for 15–20 minutes.

To make the sauce, take 500 ml of the truffle juice and 500 ml good brown stock and reduce by half in a large pan. Add 100 ml Madeira and 2 tbsp chopped truffle. Season with salt and reserve until needed. If the sauce is too thin, you can thicken it by adding a little arrowroot mixed with Madeira.

Diver's Scallop on Soba Noodles and Soy Truffle Broth with Oregon White Winter Truffles

The chef David Work (www.fiddlehead.smugmug.com) has provided the following two mouth-watering recipes, which he created for a special issue of *Fungi* magazine and tested at an epic dinner party.

Element 1: Noodles. Cook soba noodles in boiling water. When almost done, add some carrot that has been cut into long thin 'noodles' using a Japanese mandolin. Drain and rinse in warm water to stop the cooking process, but do not chill the noodles. Divide them between serving bowls in small humps.

Element 2: Broth. Begin by warming a nice rich, dark chicken stock, 2–3 fl. oz (60–90 ml) per person. Add a few drops of medium soy sauce. Add slices of Oregon winter white truffle and allow them to infuse for a few minutes. If using defrosted truffles, add some of the truffle juice from the package to the broth. Slice in some very thin shavings of garlic scapes, spring onion (scallion) or chive. Adjust the seasoning carefully.

Element 3: Scallops. I used big, beautiful sea scallops, maybe five or six of them to a pound. (We prepared them with a surprise: the night before the dinner, we cut a small slot in the bottom of each scallop and inserted a sizeable wedge of white winter truffle to infuse into the flesh overnight. I believe this made an impact on the overall experience, even if people didn't notice.) Immediately before cooking, season the scallops (one per person) with *fleur de sel* or kosher (coarse) salt and freshly ground black pepper. Sear them on both sides in a hot pan with a little olive oil or grapeseed oil, allowing the middle to remain medium rare. Place each scallop on top of the soba noodles and ladle the broth into the bowls.

Truffle Ice Cream

The base for the ice cream must be started the night before to infuse the cream with the truffles and to ensure that it is as cold as possible before churning.

14 egg yolks
¾ cup (150 g) sugar
2 cups (450 ml) milk
1 quart (0.9 litre) double (heavy) cream
5 or 6 whole Oregon black truffles

The ice cream is essentially made using a *crème anglaise* without the vanilla seeds. In a large mixing bowl, whisk the egg yolks with half of the sugar until dark yellow. Heat the milk and cream in a pan with the rest of the sugar and the whole truffles until almost simmering. Carefully temper in the hot milk mixture into the yolk mixture by adding a small amount of the hot liquid while mixing the yolks with a rubber spatula. Add a little more, still stirring. A little more. Now add the rest all at once. Place the bowl over a pan of hot water, continuing to stir gently. (This can also be done in a double boiler. The aim is not to allow the mixture to coagulate into scrambled eggs, but rather to encourage it to thicken slowly into a smooth sauce by gradually heating it.)

When you can scoop up some sauce with the spatula and drip a complete figure of eight back into the sauce before the beginning of the eight disappears, you are ready to remove the bowl to an iced-water bath to cool quickly. (You want it to reduce in temperature rapidly because you do not want the egg proteins to coagulate any further.)

Leave the truffles in the custard overnight. Before adding the custard to the ice-cream machine (following whatever directions are appropriate to your machine), remove the truffles, chop them finely and reserve them until the ice cream is ready to come out of the machine. Fold them into the ice cream while it is still soft, then put the ice cream in the freezer to firm up.

Serve the ice cream with a simple sugar cookie, and coffee topped with truffled whipped cream. For the cream, chop Oregon black truffles and add them to double (heavy) cream with sugar to taste, before whipping.

Select Bibliography

Apicius, *Cookery and Dining in Imperial Rome*, ed. Joseph
    Dommers Vehling (Mineola, NY, 1977)
Carême, Antonin, *The Royal Parisian Pastrycook and Confectioner*,
    ed. John Porter (London, 1834)
Castelvetro, Giacomo, *The Fruit, Herbs and Vegetables of Italy*
    [1614], trans. Gillian Riley (Totnes, Devon, 2012)
Ceccarelli, Alfonso, *Sui tartufi*, ed. Arnaldo Picuti and Antonio
    Carlo Ponti (Perugia, 1999)
Crosby, Alfred Jr, *The Columbian Exchange: Biological and Cultural
    Consequences of 1492* (Westport, CT, 2003)
Dedulle, Annemie, and Toni de Coninck, *Truffles: Earth's Black
    Gold* (Richmond Hill, ON, 2009)
Dubarry, Françoise, and Sabine Bucquet-Grenet, *The Little Book
    of Truffles* (Paris and London, 2001)
Dumas, Alexandre, *From Absinthe to Zest: An Alphabet for Food
    Lovers*, trans. Alan Davidson and Jane Davidson (London
    and New York, 2011)
Durante, Castore, *Herbario Nuovo* (Rome, 1585)
Flandrin, Jean-Louis, and Massimo Montanari, eds, *Food: A
    Culinary History*, trans. Albert Sonnenfeld (New York, 2000)
Fungi magazine, 1/3 (2008), special truffle issue
Giono, Jean, *The Man Who Planted Trees*, trans. Norma L.
    Goodrich (White River Junction, VT, 2007)
Hall, Ian, Gordon Brown and James Byars, *The Black Truffle: Its
    History, Uses and Cultivation* (Christchurch, New Zealand, 2001)

——, Gordon Brown and Alessandra Zambonelli, *Taming the Truffle: The History, Lore and Science of the Ultimate Mushroom* (Portland, OR, 2007)

Luard, Elisabeth, *Truffles* (London, 2006)

Maser, Chris, Andrew W. Claridge and James M. Trappe, *Trees, Truffles and Beasts: How Forests Function* (New Brunswick, NJ, 2008)

Murat, Claude, et al., 'Is the Périgord Black Truffle Threatened by an Invasive Species? We Dreaded it and it has Happened!', *New Phytologist*, CLXXVIII/4 (June 2008), pp. 699–702

Naccarato, Peter, and Kathleen LeBesco, *Culinary Capital* (London and New York, 2012)

Renowden, Gareth, *The Truffle Book* (Amberley, New Zealand, 2005)

——, 'Truffle Wars', *Gastronomica: The Journal of Food and Culture*, VIII/4 (Autumn 2008), pp. 46–50

Rittersma, Rengenier, 'Not Only a Culinary Treasure: Trufficulture as an Environmental and Agro-political Argument for Reforestation', in *Atti del Terzo Congresso Internazionale di Spoleto sul Tartufo*, ed. M. Bencivenga et al. (Spoleto, 2010), pp. 514–17

——, 'Only the Sky is the Limit of the Soil: Manifestations of Truffle Mania in Northern Europe in the 18th Century', in *Atti del Terzo Congresso Internazionale di Spoleto sul Tartufo*, ed. M. Bencivenga et al. (Spoleto, 2010), pp. 518–22

——, 'A Culinary *Captatio Benevolentiae*: The Use of the Truffle as a Promotional Gift by the Savoy Dynasty in the 18th Century', in *Royal Taste: Food, Power and Status at the European Courts after 1789*, ed. Daniëlle de Vooght (Farnham, Surrey, 2011), pp. 31–57, 202–6

——, 'Industrialized Delicacies: The Rise of the Umbrian Truffle Business (1860–1918)', *Gastronomica: The Journal for Food and Culture*, XII/3 (Autumn 2012), pp. 87–93

Safina, Rosario, and Judith Sutton, *Truffles: Ultimate Luxury, Everyday Pleasure* (Hoboken, NJ, 2002)

Sasson, Jack, 'Thoughts of Zimri-Lim', *Biblical Archaeologist*, XLVII/2 (June 1984), pp. 110–20

Trappe, Matt, Frank Evans and James Trappe, *Field Guide to North American Truffles* (Berkeley, CA, 2007)

Wang, Sunan, and Massimo Marcone, 'The Biochemistry and Biological Properties of the World's Most Expensive Underground Edible Mushroom: Truffles', *Food Research International*, XLIV (2011), pp. 2567–81

Websites and Associations

Truffle Grower and Enthusiast Associations

Australian Truffle Growers Association
www.trufflegrowers.com.au

New Zealand Truffle Association
www.nztruffles.org.nz

North American Truffle Growers' Association
www.trufflegrowers.com

North American Truffling Society
www.natruffling.org

Truffle Association of British Columbia
www.bctruffles.ca

Truffle Museums

House of Truffles and Wine, Ménerbes, France
www.vin-truffe-luberon.com

Truffle Eco-Museum, Sorges, France
www.ecomusee-truffe-sorges.com

Truffle Museum, Borgofranco sul Po, Italy
www.stradadeltartufo.org/truffle-museum

Truffle Museum, Metauten, Spain
www.museodelatrufa.com

Truffle Museum, San Giovanni d'Asso, Italy
www.museotartufo.museisenesi.org

Urbani Truffles, Scheggino, Italy
www.urbanitartufi.it

Truffle-hunting

Agriturismo Ca' Solare di Matteo Bartolini, Città di Castello,
Italy
www.agriturismo-casolare.it

La Truffe du Ventoux, Monteux, France
www.truffes ventoux.com

Vaucluse Tourism
www.provenceguide.co.uk

Truffle Fairs

Alba International White Truffle Fair, Alba, Italy
(October–November)
www.fieradeltartufo.org

Black Truffle Festival, Norcia, Italy (February)
www.neronorcia.it

Canberra and Capital Region Truffle Festival, Canberra
(June–August)
www.trufflefestival.com.au

Città di Castello, Italy (November)
www.iltartufobianco.it

Fiera Monográfica de la Trufa, Sarrión, Spain (December)
www.fitruf.es/en

Napa Truffle Festival, Napa, California, USA (January)
www.napatrufflefestival.com

Oregon Truffle Festival, Eugene (OR), USA (January)
www.oregontrufflefestival.com

Truffle Days Festival, Istria, Croatia (September–October)
www.istra.hr/en

Truffle and Wine Show, Carpentras, France (February)
www.provenceguide.co.uk

Valtopina Truffle Festival and Market, Valtopina, Italy
(November)
www.tartufoavaltopina.it

Truffle Trees and Products

American Truffle Company: 'Provider of inoculated seedlings and the science of truffle cultivation'
www.americantruffle.com

Ayme Truffe: producer of truffle trees. Also sells preserves, books, accessories and the sign 'Truffière protégée par la loi: Accés Interdit', which is 'to discourage truffles-thief'
www.ayme-truffe.com/en

Oregon Truffle Company: sellers of Oregon white truffles
www.oregontrufflecompany.com

Oregon White Truffle Oil: the only all-natural truffle oil made in the USA
www.oregontruffleoil.com

Regalis Foods: 'Purveying nature's edibles, direct from the source'
www.regalisfoods.com

Taste Arts: fresh and frozen truffles
www.tastearts.com

Truffle UK Ltd: 'The very finest truffle products' as well as inoculated seedlings
www.truffle-uk.co.uk

Truffles & Mushrooms (Consulting) Ltd: 'Involved in the cultivation of truffles and the exploitation of the beneficial effects of mycorrhizas'
www.trufflesandmushrooms.co.nz

Urbani Group: truffles and truffle food products
www.urbani.com

Acknowledgements

This book would never have been written had my colleague Simon Young not said: 'Why don't you write a proposal for Reaktion about truffles?' Thanks, Simon.

I was overwhelmed by how generous with their time a number of the world's truffle experts were. Francesco Paolocci explained the mysteries of secret truffle sex to me and showed me around his lab, and Simon Cotton helped me understand truffles' perfume. I have yet to meet Ian Hall or Matt Trappe, but they spent hours responding to my emails and shared their personal photographs for this book. Professor Yun Wang, an expert on Chinese truffles and an active proponent of Chinese environmental protection for truffle territory, was crucial for the section on Chinese truffles.

Rengenier Rittersma's work on truffles and the gastrochauvinistic competition between France and Italy was some of the most interesting scholarly work I've ever read. Rittersma is one of the world's experts (if not *the* world expert) on truffle history, and he not only shared his research with me, but also gave me input on the whole manuscript. Olga Urbani gave me access to her family's archives and answered hours of questions, and food writer Andy Ward interviewed Ian Purkayastha for me. I also learned a lot from Matteo Bartolini, truffle-hunter extraordinaire, who took me truffle-hunting a number of times with him and his dogs, Sole and Zoe. I thank him for this and all the great meals at Agriturismo Ca' Solare.

Archival images of truffles are hard to come by, but Davide Fiorino was a great help at the Accademia dei Georgofili in Florence (go and see their copy of Castore Durante's *Herbario Nuovo*). Thanks too to Alessio Assonitis, the head of the Medici Archive Project, for finding the letter from Machiavelli's son.

Many people were generous with their photographs of truffles, including the Gruppo Micologico Ternano (in particular Giorgio Materozzi and Aldobrando De Angelis). Stephen Doyle and Mauro Renna took all the photos of truffle-hunting at Ca' Solare. The ADT Vaucluse Tourisme (and Valérie Gillet in particular) bent their rules for me and shared their incredible photo galleries. The NAS Carabinieri of Bologna were allowed to talk to me about their investigations by the Comando Carabinieri per la Tutela della Salute, and Captain Simonetti kindly provided details about the NAS's blitz. Leonardo Baciarelli Falini kindly provided me with most of the more 'scientific' photos, and all photos of American truffles are by Matt Trappe and Charles Lefevre, as well as Kathleen Iudice and the American Truffle Company (which hosts the annual Napa Truffle Festival). A special thanks to David Work for his photos and recipes.

Leonardo Baciarelli Falini also suggested changes to the last two chapters, helping me more accurately to describe a truffle's life cycle. Gillian Riley and Ian Hall all gave me great feedback on the whole book, which is more solid because of their help. Writing is usually the lonely part of this process, but Bonnie Karl and Jill Edgerton encouraged me and put up with endless truffle anecdotes, as well as reading the whole manuscript. My proofreader was my best friend and copy-editor extraordinaire, John Sherck. Final thanks go to Robert Fakes from Reaktion, and to my agent, Sorche Fairbank.

The time to do this research and some of the travel necessary for it was funded by the Umbra Institute in Perugia; I am grateful to the staff (especially Mauro Renna) for all their help, as well as to the president (and my friend), Daniel Tartaglia.

Photo Acknowledgements

The author and publishers wish to express their thanks to the below sources of illustrative material and/or permission to reproduce it:

Courtesy of the Alba Truffle Festival: p. 126; photos by permission of the American Truffle Company: pp. 8, 33, 66, 116; author photos: pp. 12, 77, 133; courtesy of Biblioteca dell'Accademia dei Georgofili, Florence: p. 39; photo copyright Luca Bonacina: p. 22; reprinted from *The Man Who Planted Trees* by Jean Giono © 1985 used with permission from Chelsea Green Publishing Co., White River Junction, VT (www.chelseagreen.com): p. 83; courtesy of the Comando Carabinieri per la Tutela della Salute and the NAS of Bologna: p. 108; courtesy of Stephen Doyle: pp. 55, 134; photo courtesy Elsevier and Giovanni Pacioni: p. 11; courtesy of Leonardo Baciarelli Falini and Mattia Bencivenga: pp. 80, 97 (foot), 100, 115, 118, 121; courtesy of John Feeney/Saudi Aramco World/SAWDIA: p. 20; reproduced with the kind permission of the Italian Ministry of Cultural Patrimony and Activities, and the State Archive of Florence: p. 41; photo courtesy Jebulon/Wikimedia Commons: p. 71; photo © kcline/iStock International Inc.: p. 6; photos courtesy of Charles Lefevre: pp. 78, 95, 97 (top), 128–9; photo Library of Congress, Washington, DC: p. 16 (from the papers of John D. Whiting); collection of the artist (Pia Maria Mascaretti) reproduced courtesy of the artist: p. 124; courtesy Giorgio Materozzi, Aldobrando De Angelis and the photographic archive of the

Index

italic numbers refer to illustrations; **bold** to recipes.